SYSTEMS ENGINEERING

"The Product, the Process, and the Performance"

G. TERRY THOMAS

ISBN: 1-4392-6551-8
ISBN-13: 9781439265512

Acknowledgements

In 1958 I graduated from Drexel University with a Mechanical Engineering degree. I obtained a job at the Naval Air Development Center, in Warminster, Pennsylvania.

I later attended Temple University for three and a half years, at night, studying mathematics, and three and a half years, at night, at Penn State University, obtaining a Masters Degree in Engineering. I also attended Drexel University for one year, at night, to review my undergraduate engineering studies and obtain my Professional Engineering License.

All of these schools have provided me with an excellent engineering education, which provided me the ability to accomplish my efforts at the Department of Defense.

In 1996 the Naval Air Development Center was closed and I was moved to the Naval Air Warfare Center, in Patuxent River Maryland, and I worked until 2006.

During my 48 years of service in the Department of Defense, as a Systems Engineer and Flight Simulator Designer, many fellow engineers, military persons, and managers have provided a great deal my Systems Engineering education.

My Branch Head Manager, from 1965 to 1975, William Ogden, provided me to my initial introduction to Systems Engineering.

I would like to thank Donald Abrams for his valuable review of a portion of my book.

Enclosed in Appendix A are many of the engineers, military persons, and managers that I have worked with throughout the years.

All of these persons have provided me with some input to my Systems Engineering education.

The military makes our country safe and we must be proud and grateful to them for their work. This book provides a discussion on using a Systems Engineering approach in the generation of a product and ignores the politics of how international problems are solved.

Table of Contents

List of Figures and Tables

List of Figures

List of Tables

Chapter 1 Systems Engineering

1.1 Introduction

Systems Engineering transforms an operational need into a final product. Systems Engineering is the discipline used in the design of a product, the process development and the evaluation of the final product system. A Systems Engineer coordinates all the components of the system. The system components include the following.

- Hardware
- Software
- Modeling
- Interface
- Human Factors
- Security
- Logistics
- Safety

The Systems Engineer needs to properly coordinate each component of the system to ensure that a good product is produced.

I first became aware of the value of a Systems Engineer, in 1957, when I was a Drexel University Cooperative Student working at American Viscose Company, in Philadelphia. American Viscose was developing a new manufacturing plant to produce cellophane. The office I worked, at ninth and Market Streets, above the Gimbals Department Store, arranged their various engineering groups in separate sections throughout the office. The engineering groups included a Mechanical Engineering Section, an Electrical Engineering Section, a Plumbing Section, a Heating and Air Conditioning Section, etc.

As the new Cellophane Plant was constructed, word came back to the office that heating ducts were running into pipes and the electrical design was not compatible with the hardware requirements. There apparently was not sufficient coordination between each engineering section at our office. This is currently the job of a Systems Engineer.

Years later, my wife and I owned a television VCR with which we easily programmed many television programs to be recorded by simply typing in the television channel and recording time. After several years of use the VCR broke. Our children bought a new VCR for us, developed by a different manufacturer. The new VCR was so complex and so complicated when programming the VCR that neither my wife nor I ever used it to record a television program. I read the instructions several times, but the programming was very difficult. The VCR software may have been very well written, but obviously the human factors component was ignored.

A west coast engineer, I worked with, said that she was a System Engineer, but she was not interested in some parts of the system, such as the modeling and hardware. She was mainly interested in the system software. In my opinion, she worked on a system program, but was not a Systems Engineer.

I had a Functional group Leader who said that everything determined in a system is determined by the software.

I asked him about the selection of the instruments, displays or flight control type, of a flight simulator, and he said the type used is determined by the software engineer. In my opinion, he worked on system programs, but was not a Systems Engineer.

This book will discuss the value of a Systems Engineer and using a Systems Engineering approach in producing a good product, the process used and the importance of providing proper performance evaluation to ensure that the product meets the systems requirements.

The second section of this book will apply the Systems Engineering approach to the development of a Research and Trainer Flight Simulator and some specific details in the trainer design and development.

The International Council on Systems Engineering provides the following history of Systems Engineering.

The term System Engineering dates back to Bell Telephone Laboratories in the early 1940s (Schlager, 1956; Hall, 1962; Fagen, 1978) traces the concepts of Systems Engineering within Bell Labs back to early 1900s and describes major applications of Systems Engineering during World War II, Hall (1962) asserts that the first attempt to teach Systems Engineering as we know it today came in 1950 at MIT by Mr. Gilman, Director of Systems Engineering at Bell Labs.

Rear Admiral Grace Hopper has been quoted as saying, "Life was simple before World War II. After that we had systems."

Hall (1962) defined Systems Engineering as a function with five phases:

1. System studies or program planning.

2. Exploratory planning, which includes problem definition, selecting objectives, system Synthesis, system analysis, selecting the best system, and communicating the results.

3. Development planning, which repeats Phase 2 in more detail.

4. Studies during development, which includes the development of parts of the system and the Integration and testing of these parts.

5. Current engineering, which is what, takes place while the system is operational and being refined.

In 1990 a professional society for Systems Engineering, the National Council on Systems Engineering (NCOSE) was founded by representatives from a number of US corporations and organizations. NCOSE was created to address the need for improvements in Systems Engineering practices and education. As a result of growing involvement from Systems Engineers outside the United States, the name of the organization was changed to the International Council on systems engineering (INCOSE) in 1995. The INCOSE e-mail address is http://www.incose.org

In 1973 the University of Pennsylvania, in Philadelphia, introduced a Systems Engineering Undergraduate Program. Dr. G. Anandalingam, University of Pennsylvania, states that, "Systems Engineering considers the system as whole rather than individual components."

1.2 Systems Engineering Career History

My 48 year career was working as a Systems Engineer for the US Navy so this book and my Systems Engineering information will be related to the US Navy programs.

Since the United States earliest days, the Naval Services have fulfilled an enduring role in the defense of our national interests, according to the Naval Aviation Vision Book, published in 2003, by the Commander of Naval Air Forces. Naval Aviation, with their air wing and aircraft carriers, provided a powerful element of this mission.

The striking power of carrier based aviation provides the following enduring roles,

1. Assurance and Deterrence: A forward-deployed Navy shows our nation's commitment to allies and friends. Our forces are immediately deployable, able to project power, shape events, deter conflict, and defeat aggression.

2. Command of the Seas: Our nation's continued existence is tied to the seas; our freedom to use those seas is guaranteed by our naval forces. These combat-ready forces operate throughout the world's vast oceans to control our seaward approaches, gain and maintain control of forward Sea Lanes of Communication and keep the seas open for friendly use.

3. Power Projection: Project Naval power over great distances to deter threats. When necessary use power to disrupt, deny, or destroy hostile forces. The Naval vessels can deploy rapidly and flexibly around the globe without concern for territorial boundaries.

4. Homeland Security: Naval forces serve as a first line of defense for the American homeland, keeping attacks at bay for across the seas.

5. Deployed Forward: Detect, deter, and interdict attacks by hostile nations and emerging non-state terrorists.

Sea Power 21 encompasses three major concepts:

- Sea Strike: Projecting precise and persistent offensive power.
- Sea Shield: Projecting global defensive assurance
- Sea Bases: Projecting joint operational independence

The Naval Air Systems Command (NAVAIR) serves an important role for Sea Power 21 by providing six core technologies to Naval Aviation.

- Aircraft
- Weapons
- Sensors
- Training
- Launch and Recovery
- Communications

My Systems Engineering perspective, for this book, will use my 48 years of working experience at the Defense Department. This includes 38 years at the Naval Air Warfare Center, in Warminster Pennsylvania, from 1958 to 1996, and 10 years of working experience at the Naval Air Warfare Center, in Patuxent River Maryland, from 1996 to 2006.

The Naval Air Development Center evolved out of the Brewster Aeronautical Corporation. The Brewster

Aeronautical Corporation purchased 370 acres of farmland, in Johnsville, Pennsylvania, at a cost of $2 million dollars and began construction of a new aircraft facility, with a target completion date of July 1941. Previously, in 1938, Brewster had built two aircraft for the Navy; the 1A-1 Aircraft, a carrier based fighter, and the SBA-1Aircraft, a two seat bomber. An improved version of the fighter, the F2A-2 Aircraft was sold to England and nicknamed the "Buffalo". A small number of Buffalo Aircraft were sent to England, in the summer of 1940, during the Battle of Britain, but with armor and ammunition the Buffalo could only manage a speed of 270 mph and an altitude of 6000 ft. The projected specifications required a speed of 313 mph and an altitude of 13,000 ft. The British decided to use the World War I vintage Gladiator Bi-Plane, in place of the Buffalo Aircraft.

Shortly after Pearl Harbor, Brewster announced the development of the Buccaneer Aircraft, a new dive bomber, to be built wholly at Johnsville. The first Buccaneer Aircraft was to roll off of the production line by mid-February 1942, but production difficulties plagued the firm. When Brewster failed to deliver a single new dive bomber, President Roosevelt directed the Secretary of the Navy, Frank Knox to take immediate control of the firm. Navy Captain George Westervelt assumed command of the Brewster complex on April 21, 1942. Production problems continued until 1944 when on May 19, 1944 the Navy cancelled all contracts. The Navy took full control of the Johnsville plant, in July 1944, while Brewster moved into the manufacture of pots, pans and suitcases.

The Navy transferred part of the Naval Air Modification Unit, in Philadelphia, to the Johnsville Facility, initially with three laboratories (Pilot-less Aircraft, Aeronautical Electronics, and Armament laboratories) with 2,200 civilian employees and 450 military members, under the command of Captain Ralph Barnaby. In the fourteen months between its move to Johnsville and the surrender of Japan, the Naval Air Modification Unit repaired or experimented with over 1,370 service aircraft.

After the war, the facility concentrated on research and development; therefore the Bureau of Aeronautics renamed the facility the Naval Air Development Station and on 1 August 1949, the Naval Air Development Center. In July 1950 the Aeronautical Computer Laboratory was added to incorporate the world's largest analog computer, called the Typhoon. The Aviation Medical Acceleration Laboratory became part of the Center on June 17, 1952, when the world's largest human centrifuge was completed. The Mercury and Gemini Astronauts obtained a part of their training at this facility. The Aeronautic Instruments Laboratory (AIL) and the Aeronautical Photographic Experimental Laboratory (APEL) were transferred to NADC in December 1953, from the Naval Air Material Center, in Philadelphia. In 1958 the Aeronautical Instruments Laboratory added simulation, inertial navigation and computers to their capabilities. Also in 1958, the Anti-Submarine Warfare Laboratory was established at the NADC.

I accepted an engineering position at NADC on 19 June 1958.

The Patuxent River Naval Base history will be detailed later in this book in the Base Closure and Alignment Commission's decision to close the Naval Air Development Center, made effective in 1996.

Listed below is a history of the programs that I have worked on throughout my 48 year career,

Date	Position	Organization
2003-2006	Next Generation Threat System (NGTS) Quality Assurance and Independent Verification & Validation Technical Engineer	Systems Engineering Department Patuxent River Maryland
2001-2002	Tactical Combat Tactical System (TCTS) Software Working Group Leader	Systems Engineering Department Patuxent River Maryland
1999-2001	F-14 Trainer Engineer / Manager	Systems Engineering Department Patuxent River Maryland
1994-1998	Naval Air Training Department (PMA205) / Naval Air Warfare Center Project Coordinator & F-14 Trainer Engineer / Manager	Avionics Department Patuxent River Maryland & Warminster Pennsylvania
1986-1993	F-14 Trainer Engineer / Manager	Systems and Software Technology Department Warminster Pennsylvania
1984-1985	SH-2F Trainer Engineer / Manager	Software Computer Directorate Warminster Pennsylvania
1978-1983	F-14 Dynamic Flight Simulator & VTXTS Navy Trainer	Software Computer Directorate Warminster Pennsylvania
1969-1977	LAMPS Anti-Submarine Helicopter Flight Simulator & CH-46 Helicopter In-Flight Escape Systems Engineer	Systems Analysis and Engineering Department Warminster Pennsylvania
1965-1968	CH-53 & F-111 Flight Simulation Engineer	Aeronautical Mechanics Department Warminster Pennsylvania
1958-1965	Aerial Reconnaissance Camera Design Engineer	Aeronautical Photographic Experimental Laboratory Warminster Pennsylvania

Since most of my working experience is in the field of the general design and development of research and trainer flight simulators and the specific evaluation of flight simulator tactical environments, my examples of System Engineering design and development problems will be in these areas.

For my retirement luncheon after 48 years working for the Defense Department as a Systems Engineer and Flight Simulator Designer I presented my 10 major changes in the government work place. Presented below are these 10 major changes.

(1)   Shops Phased Out: When I went to work at the Naval Air Development Center in 1958 the base had major shops, including Machine Shops, Electrical Shops and Installation Shops. Prototypes and sometimes final products were basically produced in-house. In the mid-1960's industry convinced the government that prototypes and all final products should be design, developed and manufactured in industry. Most of the shops at the Naval Air Development Center were phased out over the next couple of years. During my first ten years at work my primary efforts were to design products which were then manufactured in our shops.

(2)   Digital Computers Introduced: In the mid 1960's the Computer Data Corporation (CDC) 6500 Digital Computer and the IBM 360 Digital Computer were purchased by the Naval Air Development Center and the digital computer age began.

(3)   Flexi-time Introduced: Flexi-time allowed a worker to arrive at work anytime from 0700 to 0900, stay eight and a half hours, including work and lunch, and then go home. Before flexi-time many workers were reprimanded for not arriving on time and at the end of the day

it took at least a half an hour to leave the base because every one left at the same time and there were just a couple of exits leading to a crowed street.

(4) Introduction to Systems Engineering: In 1965 The Naval Air Development Center new organization included a Systems Analysis and Engineering Department. My Branch Head, Bill Ogden, introduced me to the System Engineering concept.

(5) Provide ADP with all Contracts: The digital computer age brought in the requirement that an ADP document had to be signed before any contract package could be sent to industry. The ADP document purpose was to ensure that all contracts provided the correct computer requirements. The purpose for the ADP was excellent, but unfortunately in the best bureaucracy tradition, the ADP document quite often added six months to the contract negotiation cycle.

(6) Base Re-alignment and Closure (BRAC) Laws: In 1989 a BRAC Committee was formed to study the total Defense Department to determine a number of military bases that should be closed, moved or reduced. The BRAC Committee was again formed in 1991, 1993, 1995 and 2005. The BRAC Committee did a very good job in updating the overall Defense Department, but the personnel trauma was enormous.

(7)   Program Management Organization Provided: The Program Manager was made the sole manager of their Program. The Department Heads, the Division Heads and the Branch Heads, in this organization, will provide education for their personal, provide ideas to more efficiently design and develop a product and provide engineers for a specific program. The Department Heads, the Division Heads and the Branch Heads will not review any specific details of a program.

(8)   Provide Secretarial Work: In 1996 I had enough work to keep a secretary busy for a forty hour week. The organization decision was then made that engineers would have to do all of their own secretarial work. I learned Word, Power Point, and Excel and generated 100 page reports and 50 viewgraph presentations on my own. I personally really liked the change.

(9)   E-mail: In 1996 e-mail became a very favorable program element and a very valuable element. The first e-mail was actually presented in 1971, but by 1996 sending an e-mail became a daily common occurrence.

(10)   Retirement: On 19 June 2006 my retirement provided me the time to write this book.

Chapter 2 Systems Engineering Technology

2.1 Systems Engineering Technical Approach

Providing a Systems Engineering Approach in the design, development and evaluation of a product is what is required to produce a high quality product. The Systems Engineering Approach simply includes ensuring that the hardware, software, modeling, interface, human factors, security, logistics and safety elements of the products are reviewed through each phase of the program.

Systems Engineering transforms an operational need into a final product. The Systems Engineering Life Cycle Process has changed throughout the years. A generic Program Life Cycle Process is presented below, in Figure 2.2.1

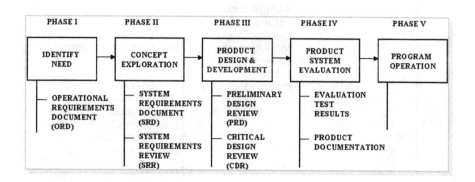

Figure 2.2.1 Program Life Cycle Process

Each of the Program Phases should address all of the Systems Engineering elements. Unfortunately, my observation is that programs usually concentrate on the software process of the program and therefore the Systems Engineering Process is secondary. A brief history of some of the Software Engineering Life Cycle Processes is presented below in Table 2.2.1, which does include many of the other Systems Engineering elements.

Table 2.2.1 Software Engineering Life Cycle Process

YEAR	PROCESS	PROGRAM TASKS	PROGRAM DOCUMENTATION
1978	DOD-STD-1679A Military Standard Defense System Software Development DOD-STD-1679A	(1) Software Performance Requirements (2) Software Design (3) Programming Standards (4) Programming Conventions (5) Software Implementation (6) Software Generation (7) Software Operation (8) Software Testing (9) Software Quality Assurance (10) Software Acceptance (11) Software Configuration Management (12) Software Management Planning	(1) Program Description Document (PDD) (2) Program Performance Specification (PPS) (3) Program Design Specification (PDS) (4) Interface Design Specification (IDS) (5) Math Model Report (MMR) (6) Data Base Design Document (DBDD) (7) Software Development Plan (SDP) (8) Program Package Document (PPD) (9) Software Quality Assurance Plan (SQAP) (10) SW Configuration Management Plan (SCMP) (11) Computer Program Test Plan (CPTP) (12) Computer Program Test Specifications (CPTS) (13) Computer Program Test Procedures (CPTP) (14) Computer Program Test Report (CPTR) (15) Software Change Proposal (SCP) (16) Computer Software Trouble Report (STR) (17) Operator's Manual (OM) (18) System's Operator's Manual (SOM)

1988	DOD-STD-2167A Military Standard Defense System Software Development DOD-STD-2167A Feb 1988	(1) Systems Requirements Analysis (2) System Design (3) Software Requirements Analysis (4) Preliminary Design (5) Detailed Design (6) Software Coding and Testing (7) Software Integration and Testing (8) System Integration and Testing	(1) System/Segment Specifications (SSS) (2) Software Product Specification (SPS) (3) Software Requirements Specification (SRS) (4) Software Development Plan (SDP) (5) Software Design Document (SDD) (6) System / Segment Design Document (SSDD) (7) Software Test Plan (STP) (8) Interface Requirements Specification (IRS) (9) Interface Design Document (IDD) (10) Software Test Description (STD) (11) Software Test Report (STR) (12) Version Description Document (VDD) (13) Software Programmer's Manual (SPM) (14) Software User's Manual (SUM) (15) Firmware Support Manual (FSM) (16) Computer Resources Integrated Document (CRID) (17) Computer System Operator Manual (CSOM)
1994	MIL-STD-498 Software Development & Documentation Military Standard Manual MIL-STD-498 December 1994	(1) Project Planning and Oversight (2) Establishing a Software Dev Environment (3) System Requirement Analysis (4) System Design (5) Software Implementation and Unit Testing (6) Unit Integration and Testing (7) CSCI Qualification Testing	(1) Software Development Plan (SDP) (2) Software Test Plan (STP) (3) Software Installation Plan (SIP) (4) Software Transition Plan (STP) (5) Operational Concept Document (OCD) (6) System/Segment Specification (SSS) (7) Interface Requirement Specification (IRS)

(Continued)

YEAR	PROCESS	PROGRAM TASKS	PROGRAM DOCUMENTATION
		(8) CSCI / HWCI Integration and Testing (9) System Qualification Testing (10) Preparing For Software Use (11) Preparing For Software Transition (12) Software Configuration Management (13) Software Product Evaluation (14) Software Quality Assurance (15) Corrective Action (16) Joint Technical and Management Reviews (17) Other Activities	(8) System/Segment Design Document (SSDD) (9) Software Requirements Specification (SRS) (10) Software Design Document (SDD) (11) Interface Design Document (IDD) (12) Database Design Description (DBDI) (13) Software Test Description (STD) (14) Software Test Report (STR) (15) Software Product Specification (SPS) (16) Software Version Description (SVD) (17) Software User Manual (SUM) (18) Software Input/Output Manual (SIOM) (19) Computer Operation Manual (COM) (20) Computer Programming Manual (CPM) (21) Firmware Support Manual (FSM)
1995	IEEE-12207 Institute of Electrical and Electronics Engineers (IEEE) Software Life Cycle Standards 1995	(1) System Requirements Analysis (2) System Design (3) Software Requirements Analysis (4) Software Design (5) Software Code & Unit Test (6) Software Unit Integration (7) Software Item (8) Software / Hardware Integration (9) System Qualification Testing	(1) Operational Concept Document (OCD) (2) System/Segment Specification (SSS) (3) Interface Requirement Specification (IRS) (4) System/Segment Design Document (SSDD) (5) Interface Design Components (IDC) (6) Software Requirements Specification (SRS) (7) Software Design Document (SDD) (8) Interface Design Document (IDD) (9) Database Design Description (DBDI) (10) Software Test Description (STD) (11) Software Test Report (STR)

1995	CMM LEVEL 2 & 3 The Capability Maturity Model Book Carnegie Mellon University 1995	(1) Requirement Management (2) Software Project Planning (3) Software Project Tracking & Oversight (4) Software Subcontract Management (5) Software quality Assurance (6) Software Configuration Management (7) Organization Process Focus (8) Organization Process Definition (9) Training Program (10) Integrated Software Management (11) Software Product Engineering (12) Inter-Group Coordination (13) Peer Review	(1) System Requirement Document (SRD) (2) Software Development Plan (SDP) (3) Configuration Management Plan (CMP) (4) Software Quality Assurance Plan (SQAP) (5) Project Training Plan (PTP) (6) Software Design Document (SDD) (7) Interface Design Document (IDD) (8) Requirements Traceability Matrix (RTM) (9) Users Manual Program documents needed but not specifically CMM Level 2 & 3 Documents (1) System Test Plan (STP) (2) System Test Procedure (STP) (3) Acceptance Test report (ATR) (4) Validation Test Report (VTR)

The documents in Table 2.2.1 are reviewed and selected on the basis of the specific program efforts. Not all the documents are necessarily selected for a specific program and in many cases documents that are not shown should be utilized for the program.

The Systems Engineering Approach, for example, sometimes needs the Math Model Report (MMR) to provide the details of the program models. This document was identified in MIL-STD-1679A, but then ignored in subsequent system software documentation. The Systems Engineering Approach sometimes needs the Data Base Design Document (DBDD) to provide the details of the program model data. This document was identified in MIL-STD-1679A, but then ignored in some subsequent software documentation. In particular, there needs to be the required documents to clearly identify the System Engineering components.

General hardware, software, modeling, interface, human factors, security, logistics, and safety details for a general product, will be presented in Chapter 3 of this book

Specific hardware, software, modeling, interface, human factors, security, logistics, and safety details, for a specific product, will be presented in Chapters 8 through 15 for the Design and Development of a Research and Trainer Flight Simulator.

2.2 System Engineering Need Examples

The following descriptions detail some of the Product, Process and Performance Problems that have occurred during my system development efforts. Some of the problems could have been resolved using a System Engineering approach.

1. F-18 Trainer / Next Generation Threat System Program (2003-2006)

The Next Generation Threat System (NGTS) tactical environment, including aircraft models, weapon models, radar models and electronic warfare models, is used on the F-18 Trainer. The NGTS tactical environment was designed to provide a large number (3500) of aircraft models to primarily accomplish an electric warfare training mission. The NGTS aircraft model produced, only allowed 200 aircraft in a mission scenario. Also, the medium fidelity aircraft model developed did not include the capability of variable turn rates, climb rates or acceleration rates. The aircraft model was not realistic to perform the tactics of a real world aircraft, to accomplish an Air Combat Maneuvering (Dog Fight) type mission. The limitations were identified in the model performance evaluation.

Systems Engineering would help to eliminate these tactical environment problems by providing a better Systems Requirement Document (SRD). A Systems Engineering approach provides all of the details required in the product development. It is important to define the flight simulator training missions in the initial requirements, so that the components needed, to provide the mission capability, can be identified. The Systems Engineering approach must also coordinate the generation of the aircraft model with the computer computation capability.

Providing a Math Model Report (MMR) would provide complete details of the tactical environment models. The Math Model Report was not provided on the Next Generation Threat System (NGTS) Program. Also, the Next Generation Threat System (NGTS) tactical environment utilizes a data base architecture. The model equation parameter data could have been completely detailed in a Data Base Design Document (DBDD), but this was not provided. Verification of the tactical environment parameter data was pursued and many parameters were found to be incorrect. There was no formal document that provided the old parameter values and the updated parameter values.

The Next Generation Threat System (NGTS) Program is currently working to correct the aircraft model limitations. There is a considerable program cost involved in updating a product to eliminate any limitations, which a Systems Engineering approach could have eliminated.

2. F-14B & D Trainer Program (1986 to 2003)

A Trainer Flight Simulator is a ground based replica of the aircraft simulated.

During the development of the F-14D Aircraft Trainer Flight Simulator, the requirements detailed the need of 16 high fidelity aircraft models and 25 missiles to be in the air at any given time. The Contractor did not coordinate the aircraft and missile model size with the computer they used to process the models, and so the F-14D Trainer ended up with 4 high fidelity aircraft models and just 9 missiles in the air at any given time.

Coordination of the flight simulation models with the computer systems is a Systems Engineering approach that would have eliminated this problem.

During the Preliminary Design Review (PDR) of the F-14B Trainer, the question was asked if the aircraft models were exercised on the computer selected, to determine the processing time of the software models. It was remembered that during the F-14D Trainer development, which occurred before the F-14 B Trainer development that only four Close-In-Combat aircraft could be provided with the computer selected, although the requirements identified sixteen Close-In-Combat aircraft. The contractor provided the computer processor exercise, after the PDR, and reported at the Critical Design Review (CRD), that a different computer was selected for the model processing.

During the development of the F-14B & D Aircraft Trainers the requirements specified that the aircraft model performance should be the following, "Velocities, Altitudes, Climb Rates and Turn Rates of each aircraft shall accurately simulate the dynamics and performance of the air threats". Since there was no numeric values or tolerances specified it was very difficult to argue that some aircraft models did not meet the real world aircraft performance.

A Systems Engineer provided future trainer aircraft model requirements with specific real world aircraft performance numeric values and tolerances.

The F-14B & D Trainer Program included the Math Model Report (MMR), which was very helpful in understanding the aircraft, weapons and radar model designs. It was discovered early in the program that the low fidelity aircraft only provided three degrees of freedom (X, Y and Z positions) and did not provide the other three degrees of freedom (roll, pitch and roll) and therefore when the aircraft were within visual view the aircraft movements appeared phony. This problem was corrected after the Preliminary Design Review (PDR).

The F-14B & D Trainer Math Model Report did not include the aircraft tactics or skill level rules, even though they were requested several times during the design reviews. This situation made it very difficult to evaluate if the threat aircraft was providing realistic tactics during their operation. A contractor, at one design review asked, "Why do you need the tactics rules?" I said, "We need the aircraft tactics rules to understand the operation of the aircraft models." I later paid the contractor to provide me the high fidelity aircraft tactics rules, which turned out to be a series of IF / Then Statements. An example of the F-14B & D Trainer tactics rules are detailed below.

IF: The threat is approaching the ownship from the rear and the closing speed is greater than 200 kts

THEN: Skill Level 4 expects a threat High Yo Yo maneuver (to reduce the threat aircraft speed)
Skill Level 1 expects a threat Lead Pursuit maneuver

IF Threat is approaching ownship from the rear and the closing speed is less than 25 kts

THEN: Skill Level 4 expects a threat Low Yo Yo maneuver (to increase the threat aircraft speed)
Skill Level 1 expects a threat Lead Pursuit maneuver

IF: Ownship is approaching threat from the rear

THEN: Skill Level 4 and 1 expects a threat Climbing Hard Turn maneuver

The rules were used to evaluate the high fidelity aircraft tactics by setting up the situation and documenting the threats reaction to the situation. The low fidelity aircraft model tactics or skill level rules were never obtained, and their performance could not be evaluated. The Math Model Report should provide complete details of the program models.

During the F-14D Trainer Requirements Review the "out-the-window" visual display system brightness of 1/2 ft lamberts were seriously questioned. It was decide to stay with the 1/2 ft lamberts value, because of cost, even though I thought this value was much to low. After the F-14D Trainer was finished the Commander in charge walked into the forty foot dome and said, "Please turn on the visual display system" and an engineers voice was heard saying, "The visual display system is on". It is very important to provide reasonable requirements to provide a product that will be accepted.

During the F-14D Trainer Preliminary Design Review (PDR) it was identified that the specific tactical environment aircraft, weapons, and radar models were not selected. A shopping list of models was provided in the contract package, but the specific models wanted were not identified. During the first night of the F-14D Trainer PDR the team got together and the specific models were selected, mainly by the Fleet Project Team. Several weapon models were not provided with the F-14D Trainer product, because the detailed list was not provided during the contract negotiations. The System Requirements Document (SRD) should have identified the specific models prior to the contract package being sent to industry.

3. SH-2F Helicopter Trainer Program (1983-1986)

The SH-2F Helicopter Trainer Program provided updates to an existing flight simulator. These updates included a new computer, new flight control system, updated flight dynamics equations, and flight control system equations. During the final evaluation of the update results a dilemma was identified. The Fleet Project Team identified the tuning of the helicopter aerodynamic and flight controls that satisfied their subjective evaluation. The Patuxent River Test Team measurements were considerably different than the Fleet Project Team tuning and now the Fleet Project Team could not fly the aircraft. The two groups would not agree on the final tuning, so a decision had to be made on which groups tuning would be accepted. I decided to accept the Fleet Project Teams tuning, because the pilots could at least fly the simulated aircraft. The Patuxent River Test Team then refused to sign off on the final product, so the Fleet Project Team could not be given aircraft flight hours, while flying the flight simulator. I think that the tuning difficulty resulted in the fact that there was a one second delay from the visual cue to the motion cue.

The overall problem is a System Engineering problem because it is important to coordinate all of the components of the system.

4. Dynamic Flight Simulator (DFS) Program (1980-1983)

The Dynamic Flight Simulator (DFS) utilized the Naval Air Development Center centrifuge to construct a simulated F-14A Aircraft, to conduct aircraft spin exercises. The DFS Team was constructing the simulator when a functional group manager asked if a formal System Requirements Document was generated. Our team had just started to fabricate the flight simulator without generating any requirements. The manager insisted that we generate a requirements document before we continued with the development. Even though this set the program back six months the requirements document was very valuable in providing the proper direction for producing a good product, and being able to evaluate the product.

A Systems Engineering complete requirements document must be generated for every product. This document must include the hardware, software, modeling, interface, human factors, security, logistics and safety requirement details and how each component is related.

5. SH-2F / SH-60 Helicopter Research Program (1971-1975)

The SH-2F / SH-60 Helicopter Research Flight Simulator Program were utilized to conduct Anti-Submarine mission scenarios. One experimental design was to evaluate the utilization of typical aircraft instruments versus the instruments being simulated on a television monitor. It was determined that the pilots could operate the aircraft equally well with either of the instrument arrangements. Future aircraft, such as the F-18 Aircraft, provided the instrumentation on a cathode ray tube (CRT).

The Systems Engineering approach is used to make choices regarding the selection of specific hardware that will accomplish the proper operation.

6. Helicopter Inflight Escape System (HEPS) Program (1968-1973)

The Helicopter Inflight Escape System (HEPS) Program was one of the most important and emotional programs I was ever involved, and it was not a success. The HEPS

Program was for the design and development of an inflight escape for the CH-46 Helicopter. Basically, if a CH-46 Helicopter was about to crash, the rotor blades were jettisoned, four 40 foot dia parachutes were deployed, using four tractor rockets and opened using four spreader guns. This occurred in a one sec time. The four parachutes reduced the vertical velocity to 70 ft/sec. At an altitude of 35 ft, four 10 inch diameter reto-rockets were deployed reducing the vertical velocity to 40 ft/sec, which is a survival velocity for the 25 crewpersons. The program also included impact resistant seats.

The program was reviewed in Congress and it was decided that the technology was not ready to put this system in the CH-46 Helicopter. In addition, the weight of the safety system would reduce the mission capability. Our program team was extremely disappointed and every time I hear about a Ch-46 Helicopter crashing I think about the program. I later proposed that the four retro-rockets be deployed as a last resort for the CH-46 Helicopter crew, but this suggestion was also not accepted.

A number of engineering things were identified from this program. A CH-46 Helicopter accident study was pursued during the program and it was determined that the helicopters crashed mainly because of engine failure or flight control failure and not because the helicopter rotors were damaged.

A complete helicopter inflight escape system was modeled on the IBM 360 Computer and this exercise provided me with many ideas for the design of the inflight escape system. I modeled the exercise and a Software Engineer, Bill Poley, converted it into software code. The idea for creating the methodology for providing a tactical environment validation test procedure also evolved out of this computer program, many years later.

As a young engineer I had many creative ideas for the design and development of the helicopter inflight escape system, which I had some difficulty convincing some senior engineering persons to take seriously. I later thought when I became the Program Manager of future programs that I must listen to the ideas of all young engineers, because they might be a real value to the program.

Chapter 3 Systems Engineering Product

3.1 Introduction

Systems Engineering transforms an operational need into a final product. The need is usually identified by a group of people that are very familiar with the operation of a system in question. In the Department of Defense these people are usually the military that best understand the operation of a system and the current limitations of the system. The need and the general system that will solve the need are then documented. In the government this document is currently the Operational Requirements Document (ORD), which is usually generated by the technical management, such as NAVAIR, in the Navy. The Operational Requirements Document (ORD) report outline is presented in Section 3.2 of this book. The Operational Requirements Document (ORD) is then expanded to provide the technical details in the System Requirements Documents (SRD). The System Requirements Document (SRD) report outline is presented in Section 3.3 of this book.

The ORD and SRD should include the details of the hardware, software, modeling, interface, human factors, security, logistics and safety.

The development of any product involves a tradeoff of the product cost, product schedule and the product quality. The estimate of the product cost, schedule and quality of the product is very difficult to determine and is usually based on the past experience of the development of past similar products. The better the initial product analysis and the generation of the product requirements the more accurate the product cost, schedule and quality will be. As the product development progresses the cost, schedule and quality of the product usually needs to be updated to provide the true values of each factor.

The major factor with all programs is funding. When you are out of money the program ends.

In a low bidder wins contract, the winning contractor usually is searching for more funding to complete the product the day after the contract is signed. The contractor quite often ends up in litigation over the true cost of producing the product.

Many managers are more concerned with the cost and schedule of the product, while engineers are more concerned with the quality of the product. Program awards are usually based on meeting the program cost and schedule than the quality of the product. Many managers say that only two of the three factors can be obtained, but in my opinion, the cost, schedule, and the quality of the product should be analyzed, in a System Engineering approach, and tradeoffs pursued to obtain the best product possible, with the trade-off of the three factors.

Program managers are continually trying to find a way to produce a product for less money and in a shorter schedule.

The current government approach is titled "Air Speed" which attempts to provide the following goals.

- Reducing the Cost of Doing Business
- Improving Productivity
- Increasing Customer Satisfaction

The "Air Speed" Process Tools include the following software systems.

- Lean: Eliminates waste and streamlines the number of steps in workflow processes.
- Six Sigma: Uses statistical analysis to eliminate variation between what we deliver and what the customer expects and therefore reduces time spent on re-work.
- Theory of Constraints (TOC): Eliminates process constraints so the workforce can focus on efficient operations.
- The "Air Speed" Process asks the following product questions.

The following Sigma Six questions are detailed below.

1. Is the project important? Has the project been chosen because it is in alignment with NAVAIR goals and the strategic direction of the business?

2. What is the problem statement, detailing:
   a. What is the problem?
   b. When was the problem first seen?
   c. Where is it seen?
   d. What is the magnitude of the problem? Is the problem measured in terms of quality? cycle time? Or cost efficiency not expected financial benefits? Ensure there is no mention or assumptions about causes and solutions.

3. Does a goal statement exist that defines the results expected to be achieved by the process, with reasonable and measurable targets? Is the goal developed for the "what" in the problem statement,

thus measured in terms of quality, cycle time or cost efficiency?

4. Does a financial business case exist, explaining the potential impact of the project on the Fleet, PMA's, Competencies, Budgets, or NOR?

5. Is the project scope reasonable? Have constraints and key assumptions been identified? Have Information Technology (IT) implications been addressed and coordinated with IT managers?

6. Who is on the team? Are they the right resources and has their required time commitment to the project been confirmed by competency and team?

7. What is the high level workplan? What are the key milestones?

8. Who are the customers for the process?
   a. The fleet?
   b. The PMAs?
   c. The Competencies? What are their requirements? Are they measurable? How were the requirements determined?

9. Who are the key stakeholders? How will they be involved in the project? How will progress be communicated? Do they agree with the project?

10. What kinds of barriers will need assistance to be removed? Has the development of a risk mitigation plan to deal with the identified risks been developed?

The questions presented above are important questions to be asked when any product is being developed.

The Product Factor is a Systems Engineering approach that can affect the success of providing a good product, and these questions are basically Systems Engineering questions. Systems Engineering looks at all aspects of a program and provides trade-offs with all elements, such as cost, schedule and quality of the product, including the hardware, software, modeling, interface, human factors, security, logistics and safety.

Another issue in the development of a product is Program Risk Management. Program Risk Management is a methodology to assess, track and mitigate events that might adversely impact the design and development of a product.

A risk is an undesirable situation or circumstance, which has both a probability of occurring and a potential consequence to program success. A risk event are those events that if they go wrong, the issue could result in problems in the development, production and fielding of the system. The program risk could be a technical risk, cost risk or a schedule risk. The Risk Management Process includes identification, analysis, planning, tracking and control.

A Program Management Risk Plan includes the following elements.

1. Identify and analyze program product development events.

2. Assess the likelihood of the events occurring and the consequences of the event occurring.

3. Handle the event actions to control risks.

4. Monitor progress of events toward meeting program goals.

5. Identify alternatives to achieve the proper cost, schedule and performance goals.

6. Make decisions on budget and funding priorities.

7. Provide risk information for milestone decisions.

8. Monitor the progress of the program as it proceeds.

A Program Management Risk Plan establishes a Program Risk Management Board to review the program product development risks.

3.2 Operational Requirements Document (ORD)

The Operational Requirements Document (ORD) summarizes the mission need, describes the overall mission area, describes the proposed system solution, provides a summary of the supporting analysis, discusses the specific mission the system will accomplish, the operational and support concepts, discusses the benefits of evolutionary acquisition & development and provides an estimated cost and schedule. It is important to generate the Operational Requirements Documents (ORD) with a Systems Engineering Approach by reviewing the hardware, software, modeling, interface, human factors, security, logistics and safety components during the generation of the ORD. A typical Operational Requirements Document (ORD) outline is provided below, in Table 3.2.1

Table 3.2.1 Operational Requirements Document (ORD) Outline

1. General Description of Operational Capability
   Mission Need
   Overall Mission Area Description
   Descriptions of the Proposed System
   Supporting Analysis
   Mission the Proposed System will Accomplish
   Operational and Support Concept
   Acquisition Approach
2. Threat
3. Existing System Short Falls
4. Capabilities Required
   Operational Performance Parameters
   Key Performance Parameters
   System Performance
5. Program Support
   Maintenance Planning
   Support Equipment
   Standardization, Interoperability and Commonality
   Computer Resources
   Human Systems Integration
   Logistics and Facilities Considerations
   Transportation and Basing

Geospatial Information and Services
Natural Environmental Support
6. Force Structure
7. Schedule
8. Program Affordability
Appendixes:

A. References
B. Distribution List
C. List of ORD Supporting Analysis
D. KPP Requirements

Glossary:

A. Abbreviations & Acronyms
B. Terms & Definition

This Operational Requirements Document (ORD) outline came from the following References,

- Defense Acquisition Desk book, CJCSI 3170-1B, Operational Requirements Document Generation Process, April 15, 2001.
- CJCSI6212.01B, Interoperability and Supportability of National Security Systems and Information Technology Systems
- C4ISP Guidance and Format

The Operational Requirements Document (ORD) will identify the approximate cost and schedule for producing the product.

It is also important that the technical managers decide whether the product will be produced in-house or on a contract to an outside company. This is usually determined on the basis of who is most capable of producing the final product. If it is decided to produce the product with an out-side contract, then a contract package must be generated, including a Statement of Work.

3.3 System Requirements Document (SRD)

The next phase of the product development process is to take the Operational Requirements Document (ORD) and provide it to the technical engineers to generate expanded technical details to the ORD. This is currently provided in the System Requirements Document (SRD). The System Requirements Document (SRD) provides the specific Systems Engineering details that allow the design and development team to provide the product to accomplish the need. The System Requirement Document (SRD) provides the "What you want". The Design and Development Team, an In-House Team or a Contractor Team, provides the "How to accomplish what you want". It is important to generate the Systems Requirements Document (SRD) with a Systems Engineering approach by providing the hardware, software, modeling, interface, human factors, security, logistics and safety requirements, during the generation of the SRD. It is very important that the System Requirements Document (SRD) provides specific performance specifications to ensure that the product produced has a measurable component to determine that the result is exactly what you want.

Since the Operational Requirements Document (ORD) identified the approximate cost, schedule and quality of the product, it is very important that the System Requirements Document (SRD) requirements are carefully thought out. For example, if you are generating the requirements for an aircraft trainer flight simulator, the following information needs to be considered. The training missions need to be seriously considered because they involve cost and schedule. If you want the trainer to provide an Air Combat Maneuvering (Dog Fight) capability then the "out-the-window" visual display system must have a high fidelity 360 degree "field-of-view" visual display system.

The Air Combat Maneuvering mission also requires high fidelity aircraft and weapon models, which are costly and require a higher computer processing capability. If the training mission requires a large number of aircraft models, the aircraft fidelity model will probably have less fidelity and cost less.

Detailed below, in Table 3.3.1 is a typical System Requirements Document (SRD) outline that I helped generate for the Tactical Combat Training System (TCTS) Program and the Next Generation Threat System (NGTS) Program. The Systems Engineering Approach was used for each System Requirements Document (SRD).

Table 3.3.1 System Requirements Document (SRD) Outline

1.0 Introduction
2.0 Scope
3.0 Background
4.0 System Overview
4.1 System Concept
4.2 System Description
5.0 Referenced Document
5.1 Government Documents
5.2 Non-Government Documents
5.3 Other Documents
6.0 List of Abbreviations &Acronyms
7.0 Program Requirements
7.1 Overview of Systems Functions
7.2 Hardware
7.3 Software
7.4 Modeling
7.5 Interface
7.6 Human Factors
7.7 Security
7.8 Logistics

7.9 Safety

8.0 Development Test and Evaluation

8.1 System Performance Verification &Validation

8.2 Site System Testing

8.3 Production System Testing

The Capability Maturity Model (CMM) Level 2 & 3 Documentation, detailed in Table 2.1.2, and repeated below in Table 3.3.2, identifies the documents that are utilized to develop the product, provide a design and development process, and evaluate the product performance.

Table 3.3.2 Capability Maturity Model (CMM) Level 2 & 3 Documentation Required For Each Product Design & Development Phase	
CMM Document	**Program Product Phase**
System Requirement Document (SRD)	Product Requirements
Software Development Plan (SDP) Configuration Management Plan (CMP) Software Design Document (SDD) Interface Design Document (IDD) Software Quality Assurance Plan (SQAP) Project Training Plan (PTP) Requirements Traceability Matrix (RTM) Users Manual	Product Design and Development Process
System Test Plan (STP) System Test Procedure (STP Acceptance Test report (ATR) Validation Test Report (VTR)	Product Evaluation (Not specifically part of CMM Level 2 & 3)

3.4 Systems Engineering Hardware, Software, Modeling, Interface, Human Factors, Security, Logistic, and Safety Technology

Detailed below will be general information on product hardware, software, modeling, interface, human factors, security, logistics and safety. Programs usually concentrate on the software element of the program, however if the software is incorporated into a computer that is not compatible with the software the final product will not be a success. If the software does not provide proper human factors then the product will not be user friendly and probably difficult to operate. Some of the hardware, modeling, interface, security, logistics and safety details required to ensure that the software will operate properly are presented below.

3.4.1 Hardware

3.4.1.1 Introduction

The program hardware depends on the product being developed. When I started work at the Naval Air Development Center, at Warminster, Pennsylvania, in 1958, there were no digital computers on the base. Today, very few hardware systems are developed without a digital computer. Provided below is a hardware history, a couple of hardware history biographies, the hardware products needed to design and develop a Research and Trainer Flight Simulator, and some digital computer system parameters.

3.4.1.2 Hardware History

From the website http://en.wikipedia.org/wiki/history_ of_computing_hardware.com, and an article by Mary Bellis, entitled "The History of Computers", the following hardware history is provided.

1000 BC to 500 BC: The invention of the Abacus.

1621: The invention of the slide rule.

1625: Wilhelm Schickard produced the first mechanical calculator.

1714: The invention of the type writer.

1900: John Fleming invented the vacuum tube.

1936: Konrad Zuse produced the Z1 Computer, which was the first freely programmable computer.

1942: John Atanascoff & Clifford Berry produced the ABC Computer.

1944: Howard Aiken & Grace Hopper produced the Harvard Mark I Computer.

1946: John Eckert & John Mauchly invented the ENIAC 1 Computer.

1948: Frederic Williams & Tom Kilburn produced the Manchester Computer.

1948: John Bardeen, Walter Brattain & William Shockley invented the transistor.

1949: The first computer assembler was invented.

1951: John Eckert & John Mauchly produced the UNIVAC Computer.

1953: IBM produced the 701 EDPM Computers.

1958: Jack Kirby & Robert Noyce invented the integrated circuit.

1964: Douglas Engelbart invented the computer mouse.

1965: The Digital Equipment Corporation produced the PDP-8 Computer.

1969: The initial internet was generated.

1970: The Intel Company invented the computer Read Access Memory (RAM).

1971: Faggin, Hoff & Mazor, from Intel Corporation produced the Intel 4004 Microprocessor Computer.

1971: Alan Shugart, from IBM, invented the "Floppy" Disk.

1973: Robert Metcalfe, from Xerox, created the Ethernet Computer Networking Bus.

1975: Bill Gates and Paul Allen opened Microsoft Corporation.

1976: The Apple I and Commodore Personal Computers were produced.

1976: The Z80 Microprocessor is produced.

1979: Seymour Rubenstein & Rob Barnaby created the Word Processor.

1981: The IBM Home Computer was produced.

1981: Microsoft Corporation created the MS-DOS Computer Operating System.

1983: The Apple Lisa and Macintosh Home Computers were produced.

1985: Microsoft Corporation created the Windows Computer Operating System.

3.4.1.3 Hardware History Personal Biographies

Presented below is the biography of a couple of important persons that have been involved in the design and development of computer systems and software languages.

From the website http://en.wikipedia.org/wiki/charles_ babbage.com the initial creator of the computer, Charles Babbage's biography is presented.

1. Charles Babbage (1792-1871) England: Charles Babbage's greatest claim to fame is that he didn't build the world's first computer-although he sure tried hard enough. Babbage's big mistake was being born in an age which had the basic knowledge to design such a machine, but no technology with which to build it.

In 1822, Babbage came up with a prototype model of a Difference Engine machine, which was designed to provide calculations for scientific tables. The machine was never perfected.

In 1833, Babbage came up with an Analytical Engine machine, which was designed to provide a range of calculations, using a series of punched cards. Technology of the period did not allow the Analytical Engine to be produced.

From the website http://en.wikipedia.org/wiki/grace_ hopper, the early software engineer Grace Hopper's biography is presented below.

2. Admiral Grace Hopper (1906-1992) New York: Grace Hopper graduated from Vassar in 1928 and Yale in 1934 with her Ph.D. She joined the Navy WAVES in 1941. In 1944 Grace reported to the Bureau of Ordinance Computation Project at Harvard University to work on the Mark Computing Machine. In 1949, Grace joined the Eckert-Mauchly Corporation, which later became Sperry-Rand Corporation. In 1985, Grace Hopper became Admiral Grace Hopper, in the Navy Reserves. Admiral Grace Hopper was one of the first software engineers and one of the most incisive strategic "futurists" in the world of computing. Perhaps her best known contribution to computing was the invention of the compiler, the intermediate program that translates the English language instructions into the language of the computer. Grace Hopper is said to have created the term "Computer Bug".

From the website http://en.wikipedia.org/wiki/john_ backus, the IBM creator of FORTRAN John Backus' Biography is presented.

3. John Backus (1924-) Philadelphia: John Backus' family was very wealthy. John Backus was very bright, but a poor student at Hill School, in Pottstown, Pennsylvania, and at the University of Virginia. After joining the Army in 1942, John Backus studied pre-engineering at the

University of Pittsburgh, pre-medical at Haverford College, and mathematics at Columbia University. In 1949, John Backus joined the IBM Computer Center working on the Selective Sequence Electronic Calculator (SSEC), one of IBM's early electronic computers.

In 1953, John Backus generated the outline of a programming language for IBM's new 704 Computer. IBM approved Backus's proposal and hired a team of programmers and mathematicians to work with him.

The 704 Computer had a built-in scaling factor, called a floating point and an indexer, which significantly reduced the operating time. The challenge that Backus and his team faced was not designing the language, but designing a device that would translate that language into something that the machine could understand. This device, known as a translator, would eliminate the laborious hand-coding that characterized computer programming at the time.

In the fall of 1954, Backus and his team published a paper describing the IBM Mathematical FORmula TRANslating (FORTRAN) system. FORTRAN was designed for mathematicians and scientists, and remains the preeminent programming language in these areas today. It allows people to work with their computer without having to understand how the machine actually works, and without having to learn the machines' assembly language. FORTRAN is still used 40 years after its introduction which is a testimony to Backus' vision.

The FORTRAN Compiler took two years to develop, consisting of 25,000 lines of machine code, stored on a magnetic tape.

3.4.1.4 Hardware Products

This book provides for the design and development of a Research and Trainer Flight Simulator. The hardware required for this product includes an aircraft cockpit crew station, an "out-the-window" visual display system, a real world image generator system, cockpit / computer interface hardware, and a number of digital computers. The cockpit crew station includes flight controls, instrumentation, displays, sensors, panels, switches and indicators.

Presented below are some of the flight simulator hardware choices that need to be selected when determining the specific hardware that will satisfy the requirements. Chapter 12 will provide the specific details of many of program hardware components.

Cockpit Configuration:

Generic
Exact Replica
Generic & Exact Replica Combination
Integration Lab

Aircraft Type:

Fixed Winged
Helicopter
Vertical Short Field Takeoff & Landing
  (V/STOL)

Seating Configuration:

Single Seat
Side by Side
Tandem
Three Crewpersons
Four Persons

Cockpit Instrumentation:

Airspeed Indicator
Barometric Altimeter
Radar Altimeter
Attitude Indicator
Horizontal Situation Indicator
Rate-of Climb Indicator
Accelerometer
Angle-of-Attack Indicator
Wing Sweep Indicator

Cockpit Display:

Multipurpose
Vertical Situation Display
Horizontal Situation Display
Digital Display
Tactical Information Display
Head-Up Display

Systems Engineering Product

Cockpit Flight Controls:

Control Stick
Side Arm Controller
Collective
Throttle
Rudders

Sensors:

Radar
Television
LOFAR Buoy
DIFAR Buoy
CASS Buoy
Infra-Red Search & Track System (IRST)
Magnetic Anomaly Detection System (MAD)
Electronic Support Measures System (ESM)

Computer:

Digital
Analog
General Purpose
Graphics Image Generator
Graphics Work Station

Crewstation / Computer
Interface

D / A Converter (Instruments)
A / D Converter (Flight Controls)
Indiscrete (Switches)
Outdiscrete (Lights)
Synchro Converter
Video

3.4.1.5 Hardware Computer System Parameter Analysis

Presented below, in Figure 3.4.1.5.1, is an illustration of a computer system with the parameters that are needed to be reviewed to provide software that is compatible with the computer system.

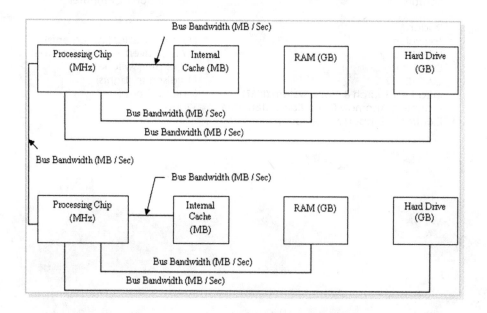

Figure 3.4.1.5.1 Computer Processing Systems Block Diagram

The processing system parameters to be reviewed including the following,

1. System Architecture

2. Number of Processors

3. Processing Capabilities (MIPS) [Million of Bits of Information Per Seconds]

4. Distribution of Software Modules between Processors

5. Internal Cache (Memory) Capability (Kilo Bytes)

6. External Cache (Memory) Capability (Mega Bytes)

7. Bandwidth between Internal Cache and Processor (Mega Bytes / Sec)

8. Read Only Memory (RAM) Capability (Giga Bytes)

9. Bandwidth between RAM and Processor (Mega Bytes / Sec)

10. Distribution of Memory Information to Internal Cache, External Cache and RAM

11. Hard Drive Capability (Giga Bytes)

12. Bandwidth between Hard Drive and Processor (Mega Bytes / Sec)

3.4.2 Software

3.4.2.1 Introduction

Software has become the key element of most program products. Software has become the most expensive element of most program products. Therefore, more emphasis is placed on the design, development and evaluation of program software than any other element a program product. Provided below is a software history, some software program structures, software operating systems, software requirements, software metrics, military software review and a software cost estimate procedure.

3.4.2.2 Software History

Software Engineering, initiated in 1968, is the discipline of designing, creating and maintaining software by applying technologies and practices from computer science, project management, engineering, application domains, interface design, and digital asset management.

From the website entitled http://www.inventors.about.com, the following history of software programming is provided.

The first generation of software codes used to program a computer was called machine language or machine code. It is the only language a computer understands. A sequence of 0s and 1s that the computer's controls interpret as instructions, electrically.

The second generation of code was called assembly language. Assembly language turns the sequences of 0s and 1s into words like 'add'. Assembly language is always translated back into machine code by programs called assemblers.

The third generation of code is called high level language, which has human sounding words and syntax (words in a sequence). The high level language was developed to allow portability between different computers. In order for the computer to understand the high order language, a compiler is used. A compiler translates the high level language into either assembly language or machine language. All software programming languages need to be translated into machine code for a computer to use the instructions they contain.

The fourth generation of software code are nonprocedural, that specify what is to be accomplished without describing how.

The fifth generation of software code are an outgrowth of artificial intelligence, is useful for programming logical processes and making deductions automatically.

From the website entitled http://en.wikipedia.org/wiki/zilog_z80, the following software language history is provided.

1946: Konrad Zues, a German engineer, develops the world's first programming language Plankalkul while hiding in the Bavarian alps waiting for World War II to end. He uses it as a chess opponent on his Z3 Computer. The language included tables and data structures.

1949: The first commonly used programming language is developed. It is to be compiled with machine code by hand.

1951:   Future Admiral Grace Hopper begins developing the compiler, which is known as AT-3 Compiler.

1952:   Alick Glennie, develops the auto-code compiler, which compiles symbolic language on a Manchester Mark I Computer.

1954:   IBM begins the development of the FORTRAN software language. The leader of the development group was John Backus.

1958:   John McCarthy begins developing the LIST software language, which is the base for artificial intelligence applications. Artificial Intelligence uses a computer to model the behavioral aspects of human reasoning and learning.

1959:   The COBOL software language is created at the Data Systems and Language Conference. COBOL is a common business software language. The COBOL software language was developed by a committee of computer manufacturers, under the leadership of Grace Hopper, a US Navy programmer.

1960:   The ALGOL 60 software language is created as the first segment built language, which is later used as the basis of the PASCAL and C languages. The ALGOL 60 software language is used primarily in mathematics and science.

1961:   Kenneth Iverson develops the APL software language, which is also suitable for mathematical applications.

1962:   The Bell Labs develop the SNOBOL software language, which is especially suitable for text and formulas.

1963: IBM begins development of the PL/I software language, which includes the best parts of ALGOL-60, FORTRAN and COBOL software languages.

1964: John Kemeny and Thomas Kurtz, two Dartmouth College professors, creates the BASIC software language and its compiler, which is a beginners' all-purpose software language, used as a teaching tool for undergraduates.

1967: The SIMULA software language is created, in Oslo, Norway, which is based on ALGOL 60 language.

1968: Niklaus Wirth begins developing the PASCAL software language, which is a more structured language, used for educational purposes.

1968: The Bell Labs creates the ALTRAN software language, which is a copy of the FORTRAN software language.

1969: The BCPL software language was developed in England. The BCPL software language is a low-level type less language which includes simple data types.

1970: Alain Colmerauer and Philippe Roussel, from France, created the PROLOG software language. The PROLOG software language is the logic-programming and artificial intelligence language.

1970: Charles Moore, an American astronomer, created FORTH software language, used in scientific and industrial control applications.

1970: Xerox creates the SMALL TALK software language, which is an object oriented language.

1972: Dennis Ritchie of Bell Laboratories creates the C software language, with its extensions, called C ++. Some experts say that the C software language combines the elegance and efficiency of the machine language with the high-level language readability and maintenance.

1975: Dr Wong creates the TINY BASIC software language, which takes only 2K of memory, and can be used on most computers.

1977: The Jean Ichbiah, Honeywell Company, Software Team created the ADA software language as part of a Department of Defense, Request for Proposal, detailing new software language requirements. ADA was named after Ada Augusta, the Charles Babbage biographer.

In summary, the major software languages developed include the following,

1. FORTRAN
2. LIST
3. COBOL
4. ALGOL 60
5. APL
6. BASIC
7. PASCAL
8. C and C ++
9. ADA

3.4.2.3 Software Programming Structures

Software can be implemented using a variety of computers, operating systems and high order programming languages. There are two primary techniques that are used for development of software programs. The Structured programming and Object-Oriented programming techniques are described below from the websites:

http://en.wikipedia.org/wiki/structured_programming

http://en.wikipedia.org/wiki/object_oriented_ programming

1. Structured Programming: A technique for organizing and coding computer programs in which a hierarchy of modules is used, with each having a single entry point and a single exit point. The control is passed downward through the structure without unconditional branches to higher levels of the structure. Three types of control flow are used: sequential, test and iteration. This technique utilizes the process of functional decomposition and data flow analysis to determine the underlying software architecture.

A low level structured program is composed of simple, hierarchical program flow structures. These are regarded as single statements, and are at the same time ways of combining simpler statements, which may be one of these structures, or primitive statements such as assignments or procedure calls. The three types of structure include the following,

- Concatenation: A sequence of statements is executed in order.
- Selection: One statements is executed depending on another statement, such as If, Then, Else.
- Repetition: A statement is executed, in a loop, until the program reaches a certain point. This is usually expressed with keywords, such as While, Repeat, For or Do....Until.

A high level structured program breaks down a larger piece of code into shorter subroutines, such as functions, procedures, methods, or blocks. The subroutines use local variables, instead of global variables. The high level structure uses small pieces of code so that code is easier to understand without having to understand the whole program at once.

2. Object-Oriented Programming: This technique utilizes the generation of separate objects. A class is used to define a group of related objects with the same characteristics. Class hierarchies are used to develop organizational structures that go from general to system-specific definitions.

Object-oriented programming may be seen as a collection of cooperating objects, as opposed to a traditional view in which a program may be seen as a collection of functions, or simply as a list of instructions to the computer. In Object-oriented programming, each object is capable of receiving messages, processing data and sending messages to other objects. Each object can be viewed as an independent little machine with a distinct role or responsibility.

Objected-oriented programming was developed as the dominant programming methodology during the mid-1980s largely due to the influence of C++, an extension of the C programming language. Its dominance was further cemented by the rising popularity of graphical user interfaces, for which object-oriented programming is well suited.

Object-oriented features have been added to many existing languages, including Ada, Basic, List, FORTRAN, Pascal and some others.

Object-oriented programming is intended to promote greater flexibility and maintainability in programming, and is widely popular in large-scale software engineering. By virtue of its strong emphasis on modularity, object-oriented code is intended to be simpler to develop and easier to understand later on, lending itself to more direct analysis, coding and understanding of complex situations and procedures than less modular programming.

Object-oriented programming basic terms are defined below, using my dog Giuseppe as the example.

Class: A class defines the abstract characteristics of a thing, such as attributes or properties and things it can do, such as behaviors or methods. For example the Class "Dog" would consist of traits shared by all dogs, such as breed, fur, color, and the ability to bark.

Object: A particular instance of class, such as my dog "Giuseppe" and Giuseppe has white fur.

Method: An objects specific ability, such as Giuseppe's ability to bark, all the time.

Inheritance: A subclass of a class. For example Giuseppe is a dog, but also a "Maltese". A subclass may also add a new method, such as Giuseppe loves cookies.

Encapsulation: Conceals the exact details of how a particular class works from objects that use its code or send messages to it. For example, the Dog class has a bark method. The code for the bark method defines exactly how the bark happens (first he inhales and then he exhales at a pitch and volume). Encapsulation does not need to know exactly how he barks. The reason for encapsulation is to prevent clients of an interface from depending on those parts of the implementation that are likely to change in the future, thereby allowing those changes to be made easily, without changes to clients.

Abstraction: Simplifying complex reality by modeling classes appropriate to the problem, and working at the most appropriate level of inheritance for a given aspect of the problem. For example, Giuseppe the dog may be treated as a dog much of the time, a Maltese when necessary to access Maltese-specific attributes or behaviors.

Polymorphism: A behavior that varies depending on the class in which the behavior is invoked. For example, if a dog is commanded to speak, he may bark, but if a pig is commanded to speak, he may squeal.

3.4.2.4 Computer Operating Systems

From the website http://en.wikipedia.org/wiki/operating_ system the following computer operating system information is provided,

An operating system is a computer program that manages the hardware and software resources of a computer. At the foundation of all system software, an operating system performs basic tasks such as controlling and allocating memory, prioritizing system requests, controlling input and output devices, facilitating networking, and managing files.

The two main computer operating systems are UNIX and Windows. The Unix Operating System was developed in the 1970's by Dennis Richie and Ken Thompson, from the Bell Labs. Today, the Unix Operating System, is used in server systems and work stations in academic and engineering environments. Microsoft developed the Windows NT Operating System, in the 1990's, which is used on more than 90% of all desktop computers.

Early operating system programs such as MS-DOS provided that only one process per CPU could be run at a time.

Modern operating system programs provide multitasking capabilities using Process Management (PM). Process Management involves computing and distributing CPU time as well as other resources. Most operating systems allow a process to be assigned a priority which affects its allocation of CPU time. Interactive operating systems also employ some level of feedback in which the task with which the user is working receives higher priority. Interrupt driven processes will normally run at a very high priority.

In many operating systems, such as Windows, there is a background process, which will run when no other process is waiting for the the CPU. Current computer architectures arrange the computer's memory in a hierchical manner, starting from the fastest registers, CPU cache, and random access memory and disk storage. An operating system's Memory Manager coordinates the use of these various types of memory by tracking which one is available, which is to be allocated and how to move data between them. This activity is usually referred to as virtual memory management, which increases the amount of memory available for each process by making the disk storage seem like main memory.

Memory Management also manages virtual addresses. If multiple processes are in memory at the same time, they must be prevented from interfering with each other's memory. This is achieved by having separate address spaces. Each process sees the whole virtual address space, typically from address 0 up to the maximum size of virtual memory, as uniquely assigned to it. The operating system maintains a page table that matches virtual addresses to physical addresses. These memory allocations are tracked so that when a process terminates, all memory used by that process can be made available for other processes.

The operating system can also write inactive memory pages to secondary storage. This process is called "paging".

Operating systems includes support of a variety of file system operations. Modern file systems are comprised of a hierarchy of directories. File system operations differ on the characters used to separate directories and case sensitivity. Microsoft Windows separates its path components with a backslash and its file names are not case sensitive. The Unix Operating System uses a forward slash and their file names generally are case sensitive.

The modern operating systems can appear on a network of another operating system and share resources such as files, printers and scanners, using wired or wireless connections. Many operating systems include some level of security for the following elements.

1. The operating system provides access to a number of resources, directly or indirectly, such as files on a local disk, privileged system calls, personal information about users, and users and services offered by the programs running on the system.

2. The operating system is capable of distinguishing between some requesters of these resources, who are authorized to access the resource, and others who are not authorized.

Internal security protects the computer resources from the programs concurrently running on the system. Processors use general purpose operating systems generally have a hardware concept of privilege. Generally less privileged programs are automatically blocked from using certain hardware instructions, such as those to read or write from external devices like disks. Instead, they have to ask the privileged program to read or write. The operating system therefore gets the chance to check the program's identity and allow or refuse the request.

External security protects the ports or numbered access points beyond the operating systems network address. Typically, the services include file sharing, print services, email, web sites, and file transfer protocols. The front line of security is hardware devices known as firewalls. A software firewall can be configured to allow or deny network traffic to or from a service or application running on the operating system.

Most modern operating systems contain Graphical User Interfaces (GUI) which allows for graphics on the computer system.

Most modern operating systems have a device driver to allow interaction with hardware devices.

A device driver is a specific type of computer software developed to allow interaction with hardware devices. Typically this constitutes an interface for communicating with the device, through the specific computer bus or communications subsystem that the hardware is connected to, providing commands to and receiving data from the device and the other end, the requisite interface to operating system and software applications. It is a specialized hardware dependent computer program and is also operating system specific. The operating system dictates how every type of device should be controlled.

The most widely used version of the Microsoft Windows family is Windows XP, released in October, 2001.

In November 2006, Microsoft Windows Vista was released, which includes architectural changes and a number of new security features, such as user account control.

3.4.2.5 Software Requirements

A typical software requirement is presented below. The software requirement is copied from the Joint Combat Training System (JTCTS) Program software requirements, with the specific reference to the JTCTS Program removed. For background information, the JTCTS Program provides an Air Combat Maneuvering (Dog Fight) training capability using a number of real aircraft, mission training area, and a simulated weapons capability.

1. The program software shall require compliance with SEI Capability Maturity Model (CMM) Level 3, or its equivalent level. If the contractor does not meet full compliance, a risk mitigation plan and schedule must be prepared that shall describe, in detail, actions that will be taken to remove deficiencies uncovered in the evaluation process and must be provided to the Program Manager for approval.

2. The program shall support an open software system architecture, using commercial standards, non-proprietary, portable software, a standard software language and architecture for all platforms, that allows updates to be pursued.

3. The program shall support the Defense Information Infrastructure (DII) Common Operating Environment (COE). DIICOE is a software infrastructure, a collection of reusable software components, a set of application program interfaces, and a series of specifications and standards for developing interoperable systems.

4. The newly developed software or any upgrade shall feature modularity in its design. To provide ease of maintenance, modules shall be designed along

functional lines, with high cohesion, and designed to minimize relationships between modules, with low coupling.

5. The computer processing speed, memory and Input (I) / Output (O) software requirements shall include 100 % spare capacity for future software growth, or a plan to provide the growth capability. The spare capability shall be measured at the highest level of processing use.

6. Software human factors must be utilized so that the man-machine interfaces shall be user-friendly, with emphasis on simple, intuitive operation. The software shall work well, be easy to learn, interact with other products, and provide ease of software corrections and future software developments. Systems controls and displays shall emphasize logical operation by personal using equipment with minimum operator training.

7. The software on different hardware components should be different; however software re-use is encouraged, where feasible.

8. The software metric processes, presented below, shall be tracked using automated software tools.

9. A Program System Safety Program, in conjunction with the Software Safety Program, shall be pursued to ensure that the program hardware, software, weapons, interfaces, and human factors, does not adversely affect the airworthiness of any air vehicles, and the safe function of weapon systems, support equipment, avionics and all other ancillary equipment associated with the program system, including the following.

- Platform modifications for accommodating installation of the program equipment shall not degrade the safety or combat integrity of the equipment.
- Safety considerations shall address the issue of effecting safe simulated weapons firing/launch and simulated expendable countermeasures deployment when the participant platform has live ordnance and expendables loaded in ready-to-fire configurations.
- Safety hazard analysis shall be executed, in accordance with MIL-STD-882D
- Protection against electrical shock hazards, use of toxic materials, mechanical equipment design and other personnel safety considerations shall be in accordance with MIL-STD-454.
- Program equipment intended for shore facilities shall meet the electrical installation requirements of NFPA-70 (National Electrical Code).

3.4.2.6 Software Metrics

A typical software metrics is presented below. Software metrics is a methodology used to ensure that the program software design and development will provide a successful product. The software metrics is copied from the Joint Combat Training System (JTCTS) Program software metrics, with the specific reference to the JTCTS Program removed. For background information, the JTCTS Program provides an Air Combat Maneuvering (Dog Fight) training capability using a number of real aircraft, mission training area, and a simulated weapons capability.

1. Software Personnel Metric Process: The Software Personnel Metric shall be used to track the ability of the contractor to maintain planned staffing levels and to maintain sufficient staffing for timely completion of the program. The Software Personnel Metric shall include the total personnel count, total experienced personnel count, personnel losses, planned total personnel level for each month of contract and the planned experienced personnel for each month of the contract.

2. Software Volatility Metric Process: The Software Volatility Metric shall be used to track changes in the number of software requirements and in the contractor's understanding of these requirements. The Software Volatility Metric shall include the total number of requirements, the cumulative number of requirement changes due to additions, deletions, and modifications of requirements, the current total numbers of open Software Action Items (SAI) and the number of non-testable requirements.

3. Configuration Management and Computer Resource Utilization Metric Process: The Configuration Management and Computer Resource Utilization Metric shall be used to identify the function and physical characteristics of the software during its life cycle and to track changes in the estimated/ actual utilization of target computer resources and provide warning if the limits of these resources are approached. The developed software sub-routines shall be balanced between processors to prevent overloads, during a large training mission exercise. The

contractor's Configuration Management Process shall provide the means through which the integrity and continuity of the software design and development is recorded and must include the following functions: Configuration System, Configuration Identification, Configuration Control, Configuration Status Accounting and Configuration Audits.

The following are details about each Configuration Management function.

- Configuration System: An automated Software Configuration Management support service shall be provided. The Systems Manager shall be able to monitor system performance using visual displays. The system shall use open database standards.
- Configuration Identification: The program Configuration Items (CI) shall follow an approved identification process. The mechanism to track and manage CI processes shall be identified. The system shall provide a functional description of all software CI applications being tracked and maintained by the system. The system shall categorize software segments by name, abbreviation, version number, hardware platform, operating system version, projected release, actual release and developer. The system shall provide Internet access to the CI data.
- Configuration Control: The system shall be able to provide the current status software CI's associated with each system release. All internal and external software interfaces shall be identified. Software documentation shall be established in the form of controlled libraries. The system shall provide a

web-based capability for authorized users to view the latest software release listing and their respective CM numbers.

- Configuration Status: For each software CI created / purchased and maintained for program operation and maintenance a record shall be established and kept current. For all CSCI's the record shall provide the Version Description Document (VDD) number and where the CSCI is installed. The system shall provide traceability of all changes from the original released CI configuration.

The system shall provide traceability of all changes from the original released CI configuration. The system shall maintain and report on the following system information.

- Product Configuration Status
- Configuration Documentation
- Current Baselines
- Historic Baselines
- Configuration Verification and Audit Status

- Configuration Audits: The system shall identify the tools and inspection equipment and test software necessary for evaluation and verification of software CI's. The system shall provide the capability to report the audit status of each software CI. The system shall be able to maintain a listing of valid software segment CI's being kept in the software library. The system shall be able to identify software CI information that identifies the discrepancies between the released software system and a valid software system. The system shall be able to identify the test procedure designed to validate the requirements of each CI. The system shall be able to identify the software discrepancy orrections.

- The Configuration Management and Computer Resource Utilization Metric shall include the Central Processor Unit (CPU) throughput for each processor, the volatile memory utilization for each processor, the non-volatile memory utilization for each processor, the bus type and bandwidth (MB / Sec) utilization for each system bus, the purpose, memory (words) and processing time (ms) for each CSU, CSU and CSCI, a state / CSC table to illustrate the system states and modes that each CSC executes, and the CSCI top-level architecture illustrated graphically.

4. Schedule Progress Metric Process: The Schedule Progress Metric shall be used to track the contractor's ability to maintain the software development schedule. As part of the final Software Development Plan (SDP), the contractor shall prepare a detailed "earned-value" plan which shall be used to monitor cost and schedule performance during the software development effort. In addition, the contractor shall perform a critical-path analysis (CP) for the software development effort. The Schedule Progress Metric shall include the budgeted work schedule (BWS), the work performed (WP), the schedule and work variance (SWV), and the critical path variance (CPV).

5. Design Progress Metric Process: The Design Progress Metric shall be used to measure the contractor's ability to maintain progress during the initial software

development phases. It tracks the development the Software Requirement Specifications (SRS's) during the system design and software requirements analysis phases, and the development of the preliminary design portion of the Software Design Documents (SDD's) during the preliminary design phase. Each of the documents culminates in a program review, including the Software Specification Review (SSR) for the SRS's and the Preliminary Design Review (PDR) for the preliminary design portion of the SDD's.

The Design Progress Metric provides visibility into the contractor's ability to hold the SRR and PDR on schedule. The Design Progress Metric includes the actual number of System/Segment Design Document (SDD) software requirements documented in each SRS, the actual number of SRS requirements documented as Computer Software Components (CSC) in the SDD's, the estimated number of SDD software requirements to be documented each month and the estimated number of SRS requirements to be documented as CSC's in the SDD's each month.

6. Computer Software Unit (CSU) Development Progress Metric Process: The Computer Software Unit (CSU) Development Progress Metric shall be used to track the contractor's ability to keep the CSU design, coding, unit testing, and integration activities on schedule. The Computer Software Unit (CSU) Development Progress Metric includes the identification of the specific CSU's that are designed coded, unit tested, and integrated, the total number of CSU's designed, the total number of CSU's coded, the total number of CSU's unit tested and the total number of CSU's integrated.

7. System Testing Progress Metric Process: The System Testing Progress Metric shall be used to track the

contractor's ability to maintain the CSC and CSCI testing progress and the degree to which the software is meeting systems requirements. The System Testing Progress Metric includes the identification of the specific CSU's that are integrated and the CSC's and CSCI's that are system tested, the total number of CSC tests completed, the total number of CSCI tests completed, the total number of Software Problem Reports (SPR's) generated, and the total number of open SPR's.

8. Software Size Metric Process: The Software Size Metric shall be used to track changes in the magnitude of the software development effort. The Software Size Metric shall include the identification of the specific Source Lines of Code (SLOC) of the CSU's, CSC's and CSCI's, the estimated and actual SLOC of the new CSCI's, the reused CSCI's, the modified CSCI's, the deleted CSCI's and the total CSCI's. Software Lines of Code (SLOC) shall include the lines-of-code that are executable and the total line-of-code, including comments.

9. Software Complexity Metric Process: The Software Complexity Metrics shall be used to assess the software complexity of each CSC and CSCI and possibly recommend software redesign that will reduce a training application problem. The Software Complexity Metric shall include the identification of the estimated and specific CSC and CSCI cyclomatic complexity, or equivalent.

10. Risk Management Metric Process: The Risk Management Metric shall be used to determine if there is a software problem that might prevent or delay the proper operation of the program software. A Risk Management Engineer shall identify software routines or interfaces that might prevent or delay the proper operation of the program software.

Risk priorities shall be assigned to the problems identified. A Risk Management Mitigation Plan shall be generated and executed. The software shall be monitored to ensure that the software will not cause a life cycle operations problem. The Risk Management Metric shall include the identification of CSU's, CSC's or CSCI's that might cause a risk management problem, identification of the purpose, potential risk management problem, identification of a risk priority of each CSU, CSC, and CSCI, and the generation and execution of the Risk Management Mitigation Plan.

11. Software Safety Metric Process: The Software Safety Metric shall be used to determine if there are hardware, weapons, interfaces, human factors and associated software systems that might cause hazards that could potentially lead to a mishap. This metric will also be utilized to analyze the potential program system safety problems.

A Software Safety Engineer, in conjunction with a System Safety Engineer, shall identify the hardware, weapons, interfaces, human factors and associated software systems that might cause hazards that could potentially lead to a mishap. Once identified and agreed upon as safety-critical problem, the high-risk functions must be analyzed. The system / software design that supports those high-risk functions shall be analyzed to ensure the presence of mitigating factors that reduce the probability of occurrence and the severity of the consequence to an acceptable level. In addition to validating the requirements from an airworthiness and weapons safety perspective, the software safety program shall ensure that the software implementation adequately verifies the safety critical requirements.

To ensure the safety of the software, an adequate level of quality shall be built into the software. For safety-critical software, the software development, test, and maintenance practices and processes, including quality assurance and configuration management shall employ measures that increase confidence in the safety of the software implementation.

Risk priorities shall be assigned to the problem identified. A Safety Mitigation Plan must be generated and executed. The software shall be monitored to ensure that the software will not cause a safety hazard. Software safety audits shall be conducted periodically to assess the software safety problems. Problem areas shall be identified at the audits and a Discrepancy Report (DR) generated. The Software Safety Metric shall include the identification of the CSU's, CSC's or CSCI's that might Cause a safety hazard, identification of the purpose, potential safety problem, identification of the risk priority of each CSU, CSC, and CSCI and the generation and execution of the Software Safety Mitigation Plan.

12. System Safety Metric Process: The System Requirements Document (SRD) shall provide a Software Safety Metric Process that shall be used during the product development to determine if there are hardware, weapons, interfaces, human factors and associated software systems that might cause hazards that could potentially lead to a mishap. These safety issues shall then be prioritized, analyzed, and mitigated. Listed below are program potential hazards that can be utilized by the program System / Software Safety Committee as a starting point to analyze possible program safety problems. The specific list for the specific program needs to be presented in the categories of hardware, software, weapons, interfaces and human factors.

3.4.2.7 Military Software Review

The following military software review is extracted from a report entitled "A Process Linking Safety, Threat, Availability and Other Critical Risk Causal Factors to their System-Level Mitigators" By Janet Ann Gill.

After Desert Storm, General Colin L. Powell, former Chairman of the Joint Chief of Staff, wrote the "Toolbox of Software Technology". Colin said that the information age has dawned in the armed forces of the United States: The sight of a soldier going to war with a rifle in one hand and a laptop computer in the other would have been shocking only a few years ago. Yet, that is exactly what was seen in the sands of Saudi Arabia in 1990 and 1991. In contrast to military hardware, the constantly changing arsenal of software distinguishes the United States from every other advanced military force on the globe. Software-intensive systems give the United States the technological edge to complete and win in the ever-changing, volatile world environment. When modern weapon systems are referred to as being smart, it is because software provides brains.

From a historical perspective, the design, the acquisition and the management of software-intensive systems are relatively new military endeavors. During the Vietnam War, the F-3 Aircraft used virtually no software in its weapon systems and software was used sparingly in other defense applications. Back then, software-intensive systems were characterized by big workhorse main-frames, occupying large rooms, using thousands of watts of electricity, tons of air conditioning, punched card inputs, with long overnight turnarounds. During the 1970s, the rapid evolution of sophisticated electronic circuitry provided smaller processors producing more computing power for a fraction of the cost. Technological advancements coupled to an ever-increasing demand on both hardware and software requirements, and dramatically increased the military's appetite for software intensive systems.

Presented below, in Table 3.4.2.6.1, is a history of several military aircraft and the software utilized in these aircraft

Table 3.4.2.6.1 Aircraft Software History		
Year	**Aircraft**	**Software (SLOC)**
1970	A-7E	16K
1973	EA-6B	48K
1973	F-14A	80K
1974	A-4	16K
1977	A-6E	64K
1983	F/A-18A/B	943K
1984	AV-8 B	764K
1985	EA-6B	395K
1986	F-14B	364K
1986	F/A-18C/D	2130K
1988	EA-6B	779K
1988	F-18 Night Attack	3054K
1989	AH-1	764K
1989	AV-8B Night Attack	1780K
1990	F-14D	416K
1993	AV-8B Radar	3748K
1994	F/A-18C/D XN-8	6629K
1996	F-14B	2866K
1996	AH-1 NTS	1000K
1996	F/A-18C/D SMUG/RUG	14268K
1997	EA-6B ICAP2	2203K
2001	F/A-18E/F	17101K
2007	JSF	21000K

3.4.2.8 Software Cost Estimate

Introduction

The cost estimate of the program software is a very important effort for all programs. Throughout the years there have been many software cost estimate methods documented.

The website http://sern.ucalgary.ca/courses/seng/621/ w98/hongd/report2.html provides the following information.

The seven basic steps in software cost estimation include the following.

1. Establish objectives:
2. Plan for required data and resources: Generate an early project plan.
3. Pin down the software requirements: Generate testable software requirements.
4. Work out as much detail as feasible: Provide a thorough understanding of the software job.
5. Use several independent techniques and sources: Generate a combination of techniques for software cost estimation.
6. Compare and iterate estimates: Analyze cost estimates to determine why they result in different estimates.
7. Follow-Up: Re-calculate software cost estimate as program progresses.

A number of software cost estimate methods are presented below.

1. Algorithmic Models: The methods provide one or more algorithms which produce a software cost estimate

as a function of a number of variables which relate to some software metric and cost drivers.

2. Expert Judgment: This method involves consulting one or more experts to provide the software cost estimate.
3. Analogy Estimation: This method involves reasoning by analogy with one or more completed programs to relate their actual costs to an estimate of the cost of a similar new program.
4. Parkinson's Principle: A Parkinson principle which uses the work expansion to fill the available volume.
5. Price to win: The cost estimation developed by this method is equated to the price believed necessary to win the job.
6. Top-Down Estimation: An overall cost estimate for the program is derived from global properties of the software product. The total cost is then split up among the various components.
7. Bottom-Up Estimation: Each component of the software job is separately estimated and the results aggregated to produce an estimate for the overall job.

COCOMO Model Software Cost Estimate

The construction cost model (COCOMO) is the most widely used software estimation model in the world. It was developed by Barry Boehm of TRW Company, first published in 1981. The COMOMO model predicts the effort and duration of a program based on inputs relating to the size of the resulting systems and a number of "cost drives" that affect productivity. The specific details of the COCOMO model is provided in the website presented below.

http://en.wikipedia.org/wiki/cocomo

NASA Managers Handbook Software Cost Estimate

The NASA Managers Handbook software cost estimate procedure is provided below.

Software Development Cost Estimate (C) = Design (D) + Implementation (I) + Testing (T)

$$C = D + .43 (D) + .11 (D)$$

Where D = (Lines of Code) (.31 Hours / Lines of Code) (Person Cost / Hour)

The Software Cost Estimate (C) value can vary with a number of program issues. These issues are discussed below.

Complexity Guideline: The Software Cost Estimate needs to be multiplied by the complexity value identified by one of the effort multiplier presented in Table 3.4.2.7.1

Table 3.4.2.7.1 Software Estimate Complexity Guidelines

Program Type	Environment Type	Effort Multiplier (CM)
Old	Old	1.0
Old	New	1.4
New	Old	1.4
New	New	2.3

Development Team Experience Guideline: The Software Cost Estimate needs to be multiplied by the Team Experience value identified by one of the effort multiplier in Table 3.4.2.7.2

Table 3.4.2.7.2 Software Estimate Team Experience Guidelines	
Years of Experience	Effort Multiplier (EM)
10	0.5
8	0.6
6	0.8
4	1.0
2	1.4
1	2.6

Computer Utilization: The Software Cost Estimate for the utilization of the computer (Comp) is provided in the formula presented below.

Comp = (0.0008 Hour) (Lines-of-Code) (Person Cost / Hour)

Documentation: The Software Cost Estimate for the documentation (Doc) is provided in the formula presented below.

Number of Pages = [120 + .026 (Lines-of- Code)]

Documentation (Doc) = [Number of Pages] [(4 Hours / Page) (Person Cost / Hours)]

The Total Software Cost Estimate is equal to the formula presented below.

Total Software Cost = C (CM) (EM) + Comp + Doc

3.4.3 Modeling

3.4.3.1 Introduction

A model is a combination of equations and data that provide a representation of a system being simulated. Model equations include generic equations and system-specific equations. Model architectures include data driven (equations and data separated) and a standalone architecture (equations and data together). Data driven model simulations can be thought of as having the data external to the software code. Standalone model simulations have data contained within the simulation code. Model architectures are presented below in Figure 3.4.3.1.1 The model equations are then usually converted into software code.

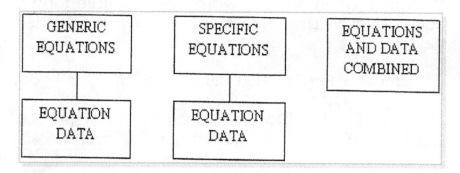

Figure 3.5.3.1.1 Model Architecture

There are an unlimited number of models that are generated for a program, depending on the program product. Research and Trainer Flight Simulator models include models of aircraft, missiles, radar, jammers, avionics, sensors, flight controls, tactics, skill levels, visual priorities, etc. Specific flight simulator modeling details are presented in Section 12.4.3 of this book. Presented below is a list of the flight simulator models detailed in Section 12.4.3

Flight Simulation Models:

Equation of Motion	Spin
Atmosphere	Carrier Bubble
Winds	Automatic Trim
Turbulence	Latitude / Longitude
Weather Systems	Magnetic Variation
IREPS	Aerodynamics
Automatic Terminal Information System (ATIS)	Weight & Balance
Carrier Control Approach (CCA)	G-Cueing
Carrier Motion	Carrier Surface Facilities
Sea State	

Aircraft Systems Models:

Air Intercept Control System (AICS)	Nose Steering
Fuel Systems	Nose Catapult
Electrical Power	Arresting Hook
Hydraulic System	Environmental Control System (ECS)
Pneumatic System	Oxygen System
Wing Sweep System	Pilot-Static System
Flap / Slats	Standby Flight Instruments
Glove Vanes	Angle of Attack (AOA) System
Speed Brakes	Canopy System
Primary Flight Controls	Ejection System
Automatic Flight Control System (AFCS)	Lighting System
Landing Gear	Control Loading System
Wheel Brake System	Control Loading Interface

Propulsion Systems Models:

Gas Generator

Engine Instruments

Bleed Air System

Engine Mode Input

Ignition System

Engine Mode Output

Starting System

Compressor Rotor Speed
Demand

Oil System

Anti-Ice & Stall

Afterburner

Engine Pressure Ratio

Variable Exhaust Nozzle
  System

Fuel Flow

Power Generation System

Nozzle Area & Thrust

Fire Detection / Extinguishing
  System

Engines Electrical System

Approach Power
  Compensator

Engine Instruments

Engine Control Systems

Air Intercept Control
System (AICS)

Communication Systems Models:

UHF Communications

Cryptic Encoder

VHF Communications

Inter-Communications System

Automatic Direction
  Finder (ADF)

Interface Blanker Unit (IBU)

IFF Transponder

On-Board Computers Models:

Computer Air
  Data Computer (CADC)

Internal Measurement
Unit (INS)

Computer System
  Data Computer (CSDC)

On Board Checkout
(OBC) Interface

Armament

Weapons Flyout

Display Interface
  Tactical Displays

Magnetic Tape
  Memory (MTM)

Navigation Systems Models:

Attitude Heading Reference System (AHRS)	TACAN
Integrated Logistics System (ILS)	Radar Altimeter
Data Link	Radar Beacon
Magnetic Compass	

Tactical Environment Models:

Object Dynamics	Scenario Management
Threat Tactics	Threat Aircraft Maneuver Logic
Threat Countermeasures	Relative Kinematics
Library Management	Threat Aircraft Dynamics

Sensor System Models:

Radar	Television Camera System (TCS)
IFF Interrogator	Tactical Air Reconnaissance Pod System (TARPS)

Electronic Warfare Models:

Jammers	Flares
Chaff	

From a "Classes of Simulation Architecture" report, by Jon Kern of Veda Corporation, dated 1994 the following model information is provided, in the next couple of pages.

A model design includes fidelity, accuracy and coupling. These terms are defined below.

Fidelity: The degree to which the model simulation fully represents all of the degrees of freedom of the physical system. For example, a 6 Degree of Freedom (DOF) system modeled as a 3 Degree of Freedom (DOF) point mass would have lower fidelity than a 6 Degree of Freedom (DOF) model.

Accuracy: The level of how faithfully a model simulation replicates the real world system that is being modeled. For example, an aircraft model simulation that matched flight tests results within +/- 5% tolerance across the entire flight envelope is more accurate than a model simulation that is within + / - 10%

Coupling: A measure of the interdependence among modules in a computer program. Low coupling, which is desired, exists when one module is not very dependent on another module. If modules pass homogeneous data items between each other, you have data coupling. High coupling, which is not desired, is exhibited in code that makes GO TO software commands to jump inside of another software routines, or that share global variables.

In addition it is important to understand if the model software code is executed in real time or non-real time; the later is also referred to as batch code.

3.4.3.2 Modeling Techniques

Systems can be modeled using different modeling methods that vary in the degree of fidelity, including the following.

1. Simple Performance Tables: Certain systems can be adequately modeled by a table of performance characteristics.

For example, a simple representation of a radar system can be developed by providing a table of characteristics such as,

- Range (Minimum and Maximum)
- Angle (Azimuth and Elevation)
- Frequency
- Pulse Repetition Frequency (PRF)

2. Transfer Function: This modeling technique provides a simplified equation, or set of equations, used to define the response of an output variable as a function of the input variable.

3. N-Degree of Freedom (DOF) Equations with Fixed Coefficients: Dynamic devices are commonly modeled as a system of equations. A 3-DOF aircraft model can be thought of as modeling a point-mass. The X, Y and Z positions are known, but the pitch, roll and yaw angles are not known. A 6-DOF aircraft model can faithfully represent both the X, Y and Z positions and the pitch, roll and yaw angles in position, velocity and acceleration. Typically, polynomial equations are built up using coefficients to represent the fact that the system is dynamic and has characteristic curves. The use of fixed coefficients represents the system at a given situation, such as a single altitude and speed. As the operating condition diverges from the design point, the model performance becomes less accurate. Fixed point modeling is often done to improve the speed of computation at the expense of accuracy, throughout the operating region.

4. N-Degree of Freedom (DOF) Equations with Variable Coefficients: The difference between this technique and the technique described in (3) above is the source of

the equation coefficients. The variable coefficients are typically derived from table look-ups that are of one or more dimensions. For example, a coefficient table may have three independent variables such as velocity, angle of attack and elevator position. The variable coefficients allow the model to be representative of actual system behavior over a wider range of operating conditions.

3.4.3.3 Model Programming Structures

The equations used for modeling a dynamic system can be implemented using a variety of computers, operating systems and high order programming languages. There are two primary techniques that are used for development of the computer situations. These techniques are described below.

1. Structured Programming: This technique utilizes the process of functional decomposition and data flow analysis to determine the underlying software architecture. Model simulation and inter-object communication or event driven programming will bog down and become ineffective. Maintenance costs tend to increase as the problem domain changes or become large and the software analysis model often represents artificial views of the world based on the programming method chosen. Programmers are more likely to fall into the traps of global variables, functions that perform too many duties, functions that are too interdependent on other modules, resulting in the hard to maintain, spaghetti code. Structured Programming for model development is not recommended by Jon Kern.

2. Object-Oriented Programming: Object-Oriented Programming models the real world as separate objects.

A class is used to define a group of related objects with the same characteristics. Class hierarchies are used to develop organizational structures that go from general to system-specific definitions. Object-oriented design strives to map the generation of representative classes and class relationships to the real world problem domain. The object-oriented analysis model serves across the entire software development cycle, from analysis to design to programming. Objects are able to communicate through the powerful abstraction of the message construct. This allows programmers to develop highly robust, stand-alone classes that are able to perform distinct services through the receipt of a message from another object. The encapsulation of object methods with object data eliminates the chance of another programmer accidentally corrupting the data. Objected-Oriented Programming for model development is recommended by Jon Kern.

3.4.3.4 Model Portability

The two major factors that enhance model simulation portability are the programming language constructs and avoidance of machine-dependent data structures. The programming language constructs used should support the majority of compilers. The avoidance of machine-dependent data structures, such as DataPool or the use of arcane techniques that rely on the machine memory storage architecture. Portability is likely higher for model simulations that are implemented as a set of equations and a set of external data sources. This allows a very clear picture of the makeup of the simulator.

3.4.4 Interface

3.4.4.1 Introduction

The number and types of interfaces are almost unlimited and complicated for the average person. Most of the interfaces are used to interface a hardware device with a computer. The main computer/device interface types are the following,

1. Serial
2. Parallel
3. SCSI
4. USB
5. ATA
6. Fire Wire (IEEE1394)

The main interface Bus Systems includes the following.

1. 1553 Bus
2. Ethernet Bus
3. VME Bus

The main interfaces between flight simulators are the following

1. Distributed Interactive Simulation (DIS)
2. High Level Architecture (HLA)

Each of the interface types, described below, are obtained from the website http://en.wikipedia.org/wiki/ with the specific interface inserted, such as, serial_interface, parallel_interface, VMEbus, SCSI, USB, Ethernet, etc.

3.4.4.2 Computer/Device Interface Systems

Presented below are the technical details for a number of the computer/device interface systems.

1. Serial Interface: Process of sending data one bit at one time, over a communication channel or computer bus. Serial communications are used for all long-haul communications and most computer networks, where the cost of cable and synchronization difficulties makes other communications impractical. Serial computer buses are becoming more common as improved technology enables them to transfer data at higher speeds. A computer serial port, such as RS-232 or RS-422 is a general purpose interface that can be used for almost any device including modems, mice and printers, although most printers are connected to a parallel port.

2. Parallel Interface: Process of sending data with more than one bit simultaneously at a time. Almost all personal computers come with at least one parallel interface. In computing, a parallel port is a type of physical interface used in conjunction with a cable to connect separate peripherals in a computer system. Over a parallel port, binary information is transferred in parallel: each bit in a particular value is sent simultaneously as an electrical pulse across a separate wire. The number of wires and the type of connector on a parallel port can vary.

3. Small Computer System Interface (SCSI): A standard interface and command set for transferring data between devices on both internal and external computer buses. The SCSI interface is most commonly used for hard disks and tape storage devices, but also connects a wide range of other devices, including

scanners, printers, CD-ROM drives, CD recorders, and DVD drives. In fact, the entire SCSI standard promotes device independence, which means that theoretically SCSI can be used with any type of computer hardware. The SCSI standards include a complex set of command protocols. A number of SCSI interface systems are presented below in Table 3.4.4.3.1

Table 3.4.4.3.1 SCSI Interface Systems				
INTERFACE	BUS WIDTH	CLOCK SPEED	MAX THROUGHPUT	MAX CABLE LENGTH
SCSI	8-bit	5 MHz	5 MB/sec	6 m
Fast SCSI	8-bit	10 MHz	10 MB/sec	1.5-3 m
Fast Wide SCSI	16-bit	10 MHz	20 MB/sec	1.5-3 m
Ultra SCSI	8-bit	20 MHz	20 MB/sec	1.5-3 m
Ultra Wide SCSI	16-bit	20 MHz	40 MB/sec	1.5-3 m
Ultra2 SCSI	8-bit	40 MHz	40 MB/sec	12 m
Ultra2 Wide SCSI	16-bit	40 MHz	80 MB/sec	12 m
Ulyra3 SCSI	16-bit	40 MHz	160 MB/sec	12 m
Ultra-320 SCSI	16-bit	320 MB/s	320 MB/sec	12 m
Ultra-640 SCSI	16-bit	640 MB/s	640 MB/sec	Unknown
SSA	1-bit	40 MB/sec	40 MB/sec	25 m
SSA 40	1-bit	80 MB/sec	80 MB/sec	25 m
FC-AL 1Gb	1-bit	100 MB/sec	100 MB/sec	500 m to 3 km
FC-AL 2Gb	1-bit	200 MB/sec	200 MB/sec	500 m to 3 km
FC-AL 4Gb	1-bit	400 MB/sec	400 MB/sec	500 m to 3 km
SAS 3 Gbit/sec	1-bit	300 MB/sec	300 MB/sec	6 m

SCSI has been commonly used in the Apple Macintosh Computer and the Sun Microsystems Computer.

4. Universal Serial Bus (USB) Interface: The USB human interface device is an interface that describes human interface devices such as, computer keyboards, computer mouse, game controllers and alphanumeric display devices. The host computer periodically polls

the devices' interrupt endpoint during operation. When the device has data to send it forms a report and sends it as a reply to the poll taken. The USB human interface device class can be used to describe both device and interface classes. The interface devices are also defined with subclass descriptors, used to device bootable. A bootable device meets a minimum adherence to a basic protocol and will be recognized by a computer. Each USB interface communicates with the host computer with a control pipe or an interrupt pipe.

5. ATA Interface: The ATA interface is a standard interface for connecting storage devices such as hard disks and CD-ROM drives inside personal computers. The ATA interface 16-bit cable length is 18 inches. Forty pin connectors are attached to a ribbon cable. Each cable has three connectors, one of which plugs into an adapter that interfaces with the rest of the computer system, with a maximum transfer rate of 80 MB/s.

6. Fire Wire: Fire Wire is the name of the Apple Computer for the IEEE 1394 interface. It is a personal computer serial bus interface standard, providing high speed communication. Fire Wire has replaced the Parallel SCSI interface in many applications due to lower implementation costs and a simplified, more adaptable cabling system.

Apple eliminated Fire Wire support in favor of USB on its newer iPods to space constraints and for wider compatibility. Fire Wire can connect together up to 63 peripherals in an acyclic topology. It allows peer-to-peer device Communication, such as communication between a scanner and a printer, to take place without using the system memory of the computer CPU. Fire Wire also supports multiple computer hosts per bus. It has a six wire cable that can supply up to 45 watts of power per port at up to 30 volts, allowing moderate consumption devices to operate without a separate power supply. Fire Wire can transfer data between devices at 100, 200, or 400 MB/s data rates. The Fire Wire 400, 6 pin connector cable, is 15 feet long, with a data rate of 480 MB/sec. Fire Wire 800, 9 pin connector cable, has a data rate of 800 MB/sec.

A Bus Network is a network architecture in which a set of clients are connected by way of a shared communications line, called a Bus. A true Bus network is passive. The computers on the Bus simply listen for a signal, but they are not responsible for moving the signal alone. The three major Bus Interface systems include the following.

1. 1553 Bus
2. Ethernet Bus
3. VME Bus

Each of these Bus systems will be described below.

1. 1553 Bus: The 1553 Bus is a serial data bus. It was originally designed, in 1975, for use with military avionics, such as the F-16 Aircraft, but has also become commonly used in spacecraft systems. The Mil Std 1553B Bus is illustrated below in Figure 3.4.4.3.1

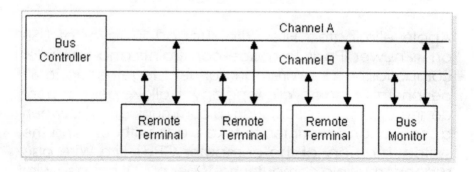

Figure 3.4.4.3.1 1553 Bus Diagram

The 1553B Bus consists of a dual-redundant bus, a bus controller, three remote terminals and a bus monitor. The 1553B Bus can provide up to 31 remote terminals. The bus controller initiates information transfers on the data bus, by sending commands to the remote terminals which reply with a response. The bus monitor is used to receive bus traffic and to extract selected information to be used at a later time. The 1553 Bus transmission rate is 1 million bits per second.

2. Ethernet: The Ethernet Bus, developed in 1975, is a twisted pair copper wire cable, defined by IEEE 802.3, used from 1990 to the present, in the most widespread local area networks (LAN). The Ethernet Bus was developed by the Xerox Corporation. The Ethernet Bus design is based on the idea of computers communicating over a shared coaxial cable acting as a broadcast transmission medium. Each Ethernet station is given a 48-bit address, which is used both to specify the destination and source of each data packet. Network interface cards do not accept packets addressed to other Ethernet stations. The Ethernet Bus does not include a bus controller. The Ethernet Bus transmission maximum rate is 10 million bits per second.

The basic Ethernet Bus system illustration and operation is presented below, in Figure 3.4.4.3.2, from the website http://en.wikipedia.org/wiki/ethernet_bus

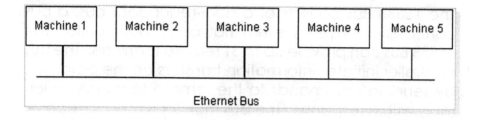

Figure 3.4.4.3.2 Ethernet Bus Diagram

The Ethernet network is connected where all machines are daisy chained using a RG58 coaxial (thin-net) cable.

Machine 2 wants to send a message to machine 4, but first it listens to make sure no one else is using the network.

If it is all clear it starts to transmit its data on to the network. Each packet of data contains the destination address, the senders address and the data to be transmitted.

The signal moves down the cable and is received by every machine on the network but because it is only addressed to number 4, the other machines ignore it.

Machine 4 then sends a message back to number 2 acknowledging receipt of the data.

When two machines transmit at the same time, a collision occurs, and each machine has to back off for a random period of time before re-trying.

3. VME Bus: The VME Bus developed in 1981 by Motorola for the 68000 Computer. The VME Bus is a scalable backplane bus interface, which includes two 32-bit buses, and two connectors, a data and address bus. The VME Bus System is illustrated in Figure 3.4.4.3.3

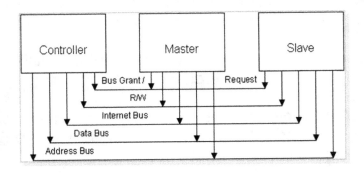

Figure 3.4.4.3.3 VME Bus Diagram

The Controller controls access to the bus and handles the interrupts. The Master takes control of the Data Bus and reads or writes data from/to Slaves. The Slave allows the Masters to read/write access and generates interrupts.

Provided below in Table 3.4.4.3.2 is a number of VME Bus systems and their capabilities.

Table 3.4.4.3.2 VME Bus List		
VME SYSTEM / SPEC	**NO OF BITS**	**TRANSMISSION RATE**
VME IEEE 1014-1987	32 Bits	40 MB/sec
VME64 VITA 1.1-1997	64 Bits	80 MB/sec
ANSI VITA 17-1998	32 Bits	160 MB/sec
VME320	64 Bits	320 MB/sec

The interface between a PC and the VME Bus is a PCI-VME 8000 card. One resides in the PCI Bus of the PC and the other resides in the VME enclosure.

Research and Trainer Flight Simulators are interfaced together through the following systems.

- Distributed Interactive Simulation (DIS)
- High Level Architecture (HLA)

Each of these interface systems will be generally described below. Specific details will be provided in the Flight Simulation part of the book.

1. Distributed Interactive Simulation (DIS): The Distributed Interactive Simulation (DIS) is a standard for conducting real-time platform-level war-gaming across multiple host computers and is used worldwide especially by military organizations, but also by other agencies such as those involved in space exploration and medicine.

The standard was developed over a series of DIS Workshops at the Interactive Networked Simulation for Training Symposium, held by the University of Central Florida's Institute for Simulation and Training. The standard itself is very closely patterned after the original SIMNET distributed interactive simulation protocol. The DIS development provided for dissimilar flight simulators, Navy and Air Force Flight Simulators, current operational equipment and testing.

Protocol Data Units (PDU) is transmitted across the DIS network. The DIS interface will transmit and receive DIS PDUs from the application to the network. This entails bundling and unbundling of the PDUs. Bundling is the ability to place more than one DIS PDU in a network data packet to increase network efficiency. The application needs to decode and encode the PDU based on the current DIS standard being used. To reduce the amount of network traffic, dead reckoning is used to estimate the position and orientation of an entity. The DIS Interface Functional Requirements Document describes what is required of a DIS interface.

2. High Level Architecture (HLA): The High Level Architecture (HLA) is a general purpose architecture for distributed computer simulation systems. Using HLA, computer simulations can communicate to other computer simulations regardless of the computing platforms. Communication between simulations is managed by a runtime infrastructure (RTI)

The High Level Architecture (HLA) consists of the following components, Interface Specification: The interface specification document defines how HLA compliant simulators interact with the Runtime Infrastructure (RTI).

The RTI provides a programming library and an application programming interface (API) compliant to the interface specification.

Object Model Template (OMT): The OMT specifies what information is communicated between simulations and how it is documented.

HLA Rules: Rules that simulation must obey to be compliant to the standards

Common terminology is used for HLA. A HLA compliant simulation is referred to as a federate. Multiple simulations connected via the RTI using a common Object Model Templates (OMT) are referred to as a federation. A collection of related data sent between simulations is referred to as an object. Objects have attributes (data fields). Events sent between simulations are referred to as interactions. Interactions have parameters (data fields).

3.4.5 Human Factors

3.4.5.1 Introduction

Leonardo Da Vinci said "Simplicity is the ultimate sophistication". Human Factors attempts to provide simplicity to the man-machine interface operation.

From the website http://www.en.wikipedia.org/wiki/human_factors.com, Human Factors is described as a term for several areas of research that include human performance, technology, design, and human computer interaction. It is a profession that focuses on how people interact with products, tools, procedures, and any processes likely to be encountered in the modern world.

Areas of interest for human factors include workload, fatigue, situational awareness, usability, user interface, learn ability, attention, vigilance, human performance, human reliability, human-computer interaction, control and display design, stress, visualization of data, individual differences, aging, accessibility, safety, shift work, virtual environments, human error, and decision making.

In Europe the term "ergonomics" is used to discuss the United States term "Human Factors".

From the website http://www.process-improvement-institute.com Human error is widely acknowledged as the major cause of quality, production, and safety risks in many industries. Although it is unlikely that human error will be completely prevented, there is a growing recognition that many human performance problems stem from a failure within organizations to develop an effective policy for managing human reliability. This is done with effective human factors planning.

My thought on human factors is the ability of a human to effectively interface with the product. Human Factors is different for different products. For example, Human Factors, dealing with aircraft hardware, requires all the elements of an aircraft cockpit, including the instrumentation, displays, flight controls, etc. to be in the exact location of the aircraft simulated, so there is no negative training, when using the hardware. Human Factors, dealing with software, is more subjective. The term "User Friendly" applies to software human factors. The software must be generated so that the human interface to the product, using the software, is intuitively utilized.

There are so many products that are not "User Friendly." For example, my wife and I were given a VCR, as a gift, that was impossible to program and after a couple of years we replaced it with a VCR that was more easily programmed. I read the instructions on the first VCR carefully, but still could not understand how to program the complex VCR. Some VCR's provide a large number of features, but the additional features and the human factors design is such that the product is difficult to operate.

3.4.5.2 Human Factors Requirements

The requirement details are provided in the MIL-STD-1472A Human Engineering Design Criteria for Military Systems, Equipment & Facilities Document, including the following.

- Achieve required performance by operator, control and maintenance personnel
- Minimize skill and personnel requirements and training time.
- Achieve required reliability of personnel-equipment combinations.
- Foster design standardization within and among systems.

3.4.5.3 Human Factor Design Examples

There is a website entitled "Bad Human Factors Designs", http://www.baddesigns.com, that provide dozens of stories about simple consumer products that are difficult to use because of a poor human factors design. Presented below are a couple of the "Bad Human Factors Designs" stories.

1. Kitchen Timer: When you set the timer to a time below 15 minutes, you must first rotate the dial past 15 minutes and then dial back below 15 minutes. There is no indication of this on the front of the timer. This is a bad human factor design.

2. Where is a soup spoon: At a cafeteria are four boxes, each containing plastic teaspoons, soupspoons, knives and forks, all with the utensils put in upside down. There is no way to know which utensil is in which box. This is a bad human factor design.

3. Solid Doors: Many facilities have doors that are used continually, but have no window, so someone push through the door may push it into a person on the other side. My daughter received a broken hand as a waitress in a restaurant having doors to the kitchen without windows. A male waiter, carrying food, kicked the door open into my daughter's hand.

4. Path: In many locations the quickest way from a door of a building to the door of another building is directly across the grass and not on the pathway, so the pathway should try to provide the quickest route from place to place.

5. Street Signs: In Las Cruces, New Mexico there is street sign with Hillrise Cir. on one street and Hillrise Dr. on the

other street. When I lived in a Holland, Pennsylvania neighborhood, we had separate streets titled Holland Road, Upper Holland Road, Lower Holland Road, Middle Holland Road and East Holland Road. This quite often caused confusion to persons coming into the Holland area.

I would like to add to the "Bad Human Factors Designs" list with a list of my own.

1. Expressway Toll Booths: One of my main frustrations is driving on a highway with a 60 mile per hour speed limit only to be tied up in a one hour traffic jam at a toll booth system.

2. Yellow Street Light: As you approach an intersection and a yellow light appears you either slow down or speed up, depending if you think the light will stay yellow as you move through the intersection or will turn red as you move through the intersection. This decision depends of the length of time the light stays yellow, but this time varies with the specific intersection. Why not provide the same time a yellow light stays on for all intersections.

3. Car Rentals: When you rent a car it is expected that every complex feature of a car will be understood, but this is not always true. A brake release, trunk release, lights activation, gas cap release, raising the windows, adjusting the seat positions, and clock setting can sometimes be very complex. Providing automobile human factors might provide intuitive operation, but a simple pamphlet or verbal explanation, provided by the car rental, would also help.

4. Telephone: When you talk on a telephone and another person is clicking in, the method for keeping one person

on hold and talking briefly to the other person usually ends up with the original person being disconnected. My telephone at work allows you to simply press a different button to go from one call to another. This operation needs to be incorporated into cell phones.

5. Gas Station: I fill up my car at a local gas station without any issues. Once I stopped at a different gas station. After putting in my charge card, the gas station computer asked for my pin number. I turned over my charge card and typed in the number on the back of the card, but this did not work I got into the car and drove over to another gas station, because the situation was too complex, at that gas station.

6. Street Signs: Enclosed below, in Figure 3.4.5.3.1, is a picture of a street sign near my son Gary's house in Hershey, Pennsylvania. The street sign doesn't clearly provide you the information you need to turn on to Raleigh Road.

Figure 3.4.5.3.1 Hershey Street Sign

3.4.6 Security

3.4.6.1 Introduction

The United States Government classification system is established under Executive Order 13292, in 2003, which provides the system of classification, de-classification and handling of national security information

The Program Security Assurance Requirement Document (SARD) specifies functional and assurance security requirements, which are derived from the International Organization for Standardization (ISO) / International Electro- technical Commission (IEC) 15408 document.

The first effort is to determine what elements of a program require security identification. The second effort is to identify the security level of the element. Security levels include Top Secret, Secret, Confidential, and No foreign Nationals (NOFORN), which are defined below from the website.

http://en.wikipedia.org/wiki/classified_information_in_the_United_States

Top Secret: Information that would cause exceptionally grave damage to national security if disclosed to the public.

Secret: Information that would cause serious damage to national security if disclosed to the public.

Confidential: Information that would cause damage to national security if disclosed to the public.

NOFORN: Information that cannot be provided to foreign nationals.

The security definitions are somewhat ambiguous and the security documents are also somewhat ambiguous, therefore determining the classification of a program element is very difficult.

My general opinion is that performance words are not classified, but performance numbers are classified. For example, radar identified with a Track-While-Scan mode is not classified, but the radar detection of 25 miles is classified. Specific classification levels should be identified by a knowledgeable classification person.

Once the program's classified elements are identified the next program procedure is to determine how the classified elements will be protected. Protection is usually provided by separating the classified elements from the unclassified elements in all documentation, models, software and the configuration management process. Many programs simply make everything classified to avoid any classification mistakes, but many elements of a program need to be discussed in an unclassified environment, which is difficult when all documents are classified.

The following industry trade secrets information is obtained from the website http://en.wikipedia.org/wiki/trade_secret

In industry, security includes Trade Secrets. A Trade Secret is a formula, practice, process, instrument, pattern, or compilation of information of information used by business to obtain an advantage over competitors within the same industry or profession. A company can protect its confidential information through non-complete non-disclosure contracts with its employees. The law of protection of confidential information effectively allows a perpetual monopoly in secret information and it does not expire, as would a patent. The lack of formal protection, however, means that a third party is not prevented from independently duplicating and using the secret information once it is discovered. The sanctioned protection of such type of information from public disclosure is viewed as an important legal aspect by which a society protects its overall economic vitality. A company typically invests time and energy into generating information regarding refinements of process and operation. If competitors had access to the same knowledge, the first company's ability to survive or maintain its market dominance would be impaired.

3.4.6.2 Security Process

Prior to 1998, at the Naval Air Warfare Center, Patuxent River, Maryland, classified information was very carefully protected. Classified documents were kept in a classified safe. A list of every classified document in the safe was carefully maintained, and reviewed each year by the Security Office. Classified documents moved from the classified safe to another place were done with a classified transmittal to keep careful track of all classified documents. Copies made of all classified documents were carefully identified and only reproduced at the Security Office.

The rooms or buildings where classified material is stored or handled must have a facility clearance at the same level as the most sensitive material to be handled. The U. S. General Services Administration sets standards for locks and containers used for storage of classified material. Electromechanical locks are used that limit the rate at which combinations can be tried out. After so many failed attempts these locks will permanently lock requiring a locksmith to reset them. There are restrictions on how classified documents can be shipped. Top secret material must go by special courier. Secret material can be sent within the United States via registered mail, and confidential material by certified mail. Electronic transmission of classified information requires the use of National Security Agency "Type 1" approved encryption systems. Computer networks for sharing classified information are segregated by the highest sensitivity level they are allowed to be transmitted over the SIPRNET (secret network) and JWICS (Top secret-SCI network).

In 1998, these policies were changed to no longer keep careful track of classified documents. NAVAIRWARCENAC

INST 5510.1A provided the classified safe custodian with the responsibility of "tracking your classified documentation, the security personnel will not have an inventory of what is in your container". Classified documents were now copied on just about any copy machine with no accounting for a document being copied. I totally disagreed with the new policy and feel that the new policy will compromise many classified documents. I generated an e-mail of my displeasure but the e-mail was totally ignored. The average engineer does not keep close accounting of their classified documents unless they are forced too.

The old classified document system protection was much too bureaucratic, but a compromise system should be developed to provide user friendly classification protection. My suggestive protection system would provide a base single security office that keeps track of all of the bases classified documents and if a classified document is copied, it must be copied in this office. This office would determine if a specific element of a program should be classified and what level of classification should be identified.

In 1990's I presented my F-14 Aircraft Research Flight Simulator TACAIR Systems Development Facility (TSDF) to a number of admirals throughout the world. I asked our local security office if I was allowed to mention if the F-14 Aircraft carried a harm missile and they said that I should use my own judgment on the issue. I thought they should have had a more specific answer.

COMNAVREGMIDLANTINST 5510.29, dated 5 May 2006, at the Norfolk Naval Base, provided a stricter security system review and re-production capability, as detailed below from INST 5510.29 page 7 and page 12.

Page 7, Section 9c: An inventory of all classified material holdings within the command will be conducted upon change of command, relief of department head or division officer responsible for classified material, transfer of the designated custodian, or when other circumstances warrant. A report of inventory will be submitted to COMNAVREG MIDLANT Security Manager and any discrepancies noted. Listings of classified holdings will be made an enclosure to the inventory report.

Page 12, Section 13: The reproduction machine located in the Security Office Building N26, is designated as being authorized for reproduction of the classified up to and including Secret. The number of reproduced secret and confidential documents must be reported for accountability and control.

Hopefully, the NAVAIRWARCENAC at Patuxent River, Maryland will adopt this security procedure.

3.4.6.3 Security Examples

Examples of information that should be considered classified are presented below from the website http://www.fas.org/sgp/library/quist2

A 1964 DOD instruction provided more-detailed example of information that might require Secret classification.

Some of those examples are as follows:

1. A war plan or a complete plan for a future operation of war not included under Top Secret, and documents showing the disposition of our forces the unauthorized disclosure of which, standing alone, could result in actual compromise of such Secret plans.

2. Defense or other military plans not included under Top Secret or (1) above including certain development and procurement plans and programs but not necessarily including all emergency plans.

3. Specific information which alone, reveals the military capabilities or state of preparedness of the Armed Forces, but not including information the unauthorized disclosure of which could result in compromise of a Top Secret plan.

4. Information that reveals the strength of our forces engaged in hostilities; quantities or nature of their equipment; or the identity or composition of units in an active theater of operations or other geographic area where our forces are engaged in hostilities, except that mailing addresses may include organization designation. Information which reveals the strength, identity, composition, or location of units normally

requires classification as Secret in time of war. In peacetime Secret classification of information pertaining to units may be appropriate when related to war plans, estimates or deployments which involve classified information.

5. Intelligence and other information, the value of which depends upon concealing the fact that the United States possesses it, except when possession of intelligence or other special operation falling within the Top Secret classification level.

6. Particulars of scientific or research projects which incorporate new technological developments or techniques having direct military applications of vital importance to the national defense.

7. Specific details or data relating to new materials or important modifications of materials which significant military advances or new technological developments having direct military application of vital importance to the national defense.

8. Information of vital importance to the national defense concerning specific quantities of war reserves.

9. Indications of weakness, such as shortages of significant or sensitive items of equipment.

10. Examples of documents that might be classified as Secret include the following.

    • Documents containing specific design details, such as diagrammatic or descriptive, of complete basic or key equipment, apparatus, instruments, or machinery employed

_____

in a critical stage of the processing and production of end products, or the methods of manufacture of such items.

- Documents containing complete encoded flow sheets, diagrams, or reactions, including specific pressures, temperatures, voltages, rates, formulae, and other operating details not detailed in the Smyth Report, specifically related to a critical step in the preparation, processing, separation, or purification of basic feed materials, and principal end products.
- Documents containing unique nuclear, physical, and chemical characteristics of end products, and critical process materials and also details of manufacture of such materials.
- Documents showing the meaning of a name or symbol used as a code, where the code name or symbol refers to matters classifiable as Secret.
- Documents pointing out the existence of unique operational or production hazards, their nature and solution.
- Details pertaining to features of special shipping containers, routes and schedules of shipments of Secret materials.

11. Examples of information that might be classified as Confidential include the following.

- Operational and battle reports which contain information of value to the enemy.
- Intelligence reports.

- Information which indicates strength of our ground, air and naval forces in the United States and overseas areas, idea, identity or composition of units, or quality of specific items of equipment.
- Research, development, production and procurement of munitions of war.
- Performance characteristics, test data, design and production data on munitions of war.
- Operational and tactical doctrine.
- Mobilization plans.

12. Examples of documents that might be classified as Confidential include the following.

- Documents and manuals containing technical information used for training, maintenance and inspection of classified munitions of war.
- Documents containing specific design details, such as diagrammatic or descriptive, of incomplete components of basic or key equipment, apparatus, instruments, or machinery employed in a critical stage of the processing and production of end products, or the methods of manufacture of such items.
- Documents containing complete un-coded flow sheets, diagrams, or reactions, including specific pressures, temperatures, voltages, rates, formulae, and other operating details related to non-critical step in the preparation, processing, separation, or purification of basic feed materials, and principal end

products where not described in the Smyth report.

- Documents containing unique physical and chemical characteristics of special materials not pertaining to the product material or process but used to overcome operational problems.
- Documents showing the meaning of a name or code names or symbols used to refer to confidential information.
- Documents relating to special investigations, clearance, or assignment of personnel who will have knowledge of, or access to, classified information wherein adverse information is reflected.
- Details pertaining to features of special shipping containers, routes and schedules of shipments of confidential materials.

3.4.7 Logistics

3.4.7.1 Introduction

Logistics provides program supportability which includes program documentation, program material parts, spare parts, program maintenance, maintenance analysis, program testing, testing equipment, and configuration management. Presented below are details involved in each of these logistics efforts.

3.4.7.2 Logistic Documentation

The Capability Maturity Model Level 2 and 3 documentation is presented in Table 2.2.1 of the Software Engineering Life Cycle Process and repeated below, with some additions.

- System Requirement Document (SRD)
- Software Development Plan (SDP)
- Configuration Management Plan (CMP)
- Software Quality Assurance Plan (SQAP)
- Project Training Plan (PTP)
- Software Design Document (SDD)
- Interface Design Document (IDD)
- Math Model Report (MMR)
- Requirements Traceability Matrix (RTM)
- Users Manual
- System Test Plan (STP)
- System Test Procedure (STP)
- Software Test Procedure (STP)
- Acceptance Test Report (ATR)
- Validation Test Report (VTR)

Other documents may be included depending on the specific program needs.

3.4.7.3 Program Materials, Spare Parts, Maintenance and Maintenance Analysis

The complete program material list must be identified including the item name, vendor part number, federal supply code of the manufacturer, quantity required, price, and possibly the mean time between failure (MTBF), and production lead time for each item.

In addition, a program material spare parts list needs to be generated with the same information that is detailed above. Spare parts need to be supplied for material that has a low mean time between failure values.

The material mean time between failure values can sometimes be determined through maintenance analysis. I found that keeping good laboratory documentation is the best way to determine material endurance. Documenting the laboratory down time, the reason for the downtime, the procedure to solve the downtime problem, and the time required. A trouble shooting guide should be generated from laboratory documentation.

Maintenance analysis, such as those detailed in the Joint Tactical Combat Training System (JTCTS) Program document, should be pursued for the following items.

1. Preventive Maintenance Analysis: Maintenance analysis should include Calibration Requirement Analysis, Corrosion Prevention and Service/Lubrication tasks. The analysis shall include parameters, ranges, accuracy, and calibration. Servicing and lubrication tasks shall be evaluated for each engineering failure modes.

2. Corrective Maintenance Analysis: Maintenance analysis shall maintain an audit trail from each specific

engineering failure mode to the resulting corrective maintenance task.

3. Level of Repair Analysis: Maintenance analysis shall document repair requirements to determine spare part requirements.

4. Task Analysis: A number of efforts should be analyzed including the following.

   - Identify logistic support requirements for each task.
   - Identify new or critical logistic support resource requirements.
   - Identify transportability requirements.
   - Identify support requirements, which exceed established goals, thresholds or constraints.
   - Provide data to support participation in the development of design alternatives to reduce operation and support costs, optimize logistic support resource requirements, or enhance readiness.
   - Provide source data for preparation of required logistic documents, such as technical manual. Training programs, etc.
   - Determine maintenance level assignment.

Based upon the results of the supportability analysis, a final maintenance and support concept will be approved. This will then be the basis for the development of the support system document, including the following.

   - Maintenance Support Plan
   - Repairable, Subassemblies, and Maintenance Significant Items Report
   - System Safety Hazard Report

- Manpower, Personnel and Training Report
- Technical Manuals

    - Installation, set-up operation, supports equipment and maintenance instructions.
    - Commercial off-the Shelf Manuals
    - Validation Test Results Report

- Support Equipment Report

3.4.7.4 Program Testing

The program shall certify the technical accuracy and adequacy of the program products through verification and validation tests. Verification is testing the product against the design, while validation is testing the product against the real world performance values. Program testing will be completely described in the Research and Trainer Flight Simulator performance evaluation section of this book.

3.4.7.5 Program Configuration Management

Program Configuration Management is required for the program requirements, program software, and the program performance discrepancy reports. The Configuration Management operation is usually done with an industry software application. There are a number of good programs, including the following.

1. IBM Rational Clear Case (Software), Clear Quest (Performance) and Requisite Pro (Requirements).

2. Telelogic DOORS

Each of these configuration management systems has operational procedures that require understanding to properly use the system. Obtaining system training is usually required to understand the software application operation.

The Configuration Management Engineer shall provide the following efforts.

1. Generate a Configuration Management Plan and enforce the policies and procedures in the plan.

2. Train users to use the configuration management system.

3. Maintain version control of all work products.

4. Manage the change control process to track all reported problems, their resolution, and implementation status for software products under configuration control.

5. Provide status accounting reports to the Configuration Control Board (CCB), and update the status accounting databases to reflect the CCB decisions.

6. Provide a mechanism for allowing and controlling changes to products under configuration control.

7. Provide a mechanism for entering, storing and providing status information.

8. Provide system maintenance, user management, backups and recovery.

9. Support the efforts of Independent Verification and Validation (IV&V) procedures.

10. Establish library control of configuration management elements.

11. Support system testing to ensure the proper functionality of the software product prior to the release to the customer.

12. Maintain and control access to current copies of documentation and code.

13. Distribute approved configuration management elements

14. Generate Configuration Management reports.

Specific Configuration Management details will be provided in the Research and Trainer Flight Simulator development section of the book.

3.4.8 Safety

3.4.8.1 Introduction

The Safety process is designed to protect personnel from accidental death, injury or occupational illness when using weapon systems, equipment, material and facilities. A program system safety approach manages the risk of mishaps associated with program operations, through the risk management consistent with mission requirement, in the technology development by design for systems, subsystems, equipment, facilities and their interfaces and operation.

The Department of Defense, Standard Practice for System Safety, MIL -STD-882D, dated 10 February 2000, provides the approach to manage safety and health mishap risks encountered in the development , test, production, use of the DoD systems, subsystems, equipment and facilities. Mishap risks must be identified, evaluated and mitigated to a level acceptable to the appropriate authority and compliant with federal laws and regulations.

3.4.8.2 Safety Approach

From a presentation entitled "Overview of How and Why to Execute a System Safety Program" by John Jabara and Janet Gill the following Program Safety Approach is provided.

A System Safety Program determines if the hardware, software, modeling, interface, human factors, security and logistics designs adversely affect the air worthiness of air vehicles and the safe function of weapon systems, support equipment, avionics and all other aircraft ancillary equipment. The objective of the Safety Program is provide management and engineering guidelines to achieve an acceptable level of risk for the system through a systematic approach of hazard analysis, risk assessment and risk management.

The System Safety Process provides for the following efforts.

1. Establish system safety program requirements, scope, ground rules and assumptions

2. Identify safety requirements

3. Perform safety hazard analysis on all program system elements

4. Identify system safety hazard risks on all program system elements

5. Prioritize the safety hazard risks as frequent, probable, occasional, remote or improbable

6. Conduct trade studies to determine if safety hazard risks are mitigated through industry

7. Provide safety hazard risk mitigation plan

8. Implement safety hazard mitigation plan

Presented below is the abstract from a report entitled, " A Process Linking Safety, Threat, Availability and Other Critical Risk Causal Factors To Their System-Level Mitigators, Relative to Hardware, Software, Human System Integration and Cost", by Janet Ann Gill, 2004.

The focus of this work is the creation of a manual process that aids the design and safety risk analysis of military and other critical systems. The objectives of the process are to increase safety-critical and mission-critical success and gather evidence and the risk of losing life, property and damaging the environment is as low as possible. This process has the potential to identify and gather evidence to minimize the causes of safety-related design and implementation errors in software safety-critical systems and assist engineers and managers to pick the best alternative mitigators with respect to technology maturity, lifestyle operation, and cost. This process lends itself well to analyzing the safety requirements found in the "system of systems" of today, where there is a high degree of interdependence and interoperability and an increasing number of off-the-shelf components.

Now that the safety community has recognized that both software and human involvement form part of the system to be analyzed, a new goal is to identify which hardware, software and human system integration functions are safety-critical. This report describes a new process to systems engineering analysis of safety-critical software and gives examples of successful safety critical analysis where different disciplines are brought together to participate in the overarching safety process.

The solution described in this report helps to close the gap between System Safety and Software Engineering analysis activities and identify and link hazards and undesired events and their causal factors to their optimal mitigating requirements. If there is a link, mitigation has been achieved. If there is no link, a new mitigating requirement needs to be determined. This allows the software engineers to trace the safety-critical requirements to test or proof of implementation at the appropriate level of rigor and dependability.

Aside from proving mitigation, the uncovering of missing specifications or requirements is a major value added of this manual process. Over the life of this research there has been the opportunity to apply this tailorable process to over twenty-five Naval Aviation aircraft, weapon and aircraft components. A semi-automated mechanism to support the maintenance of the links from the hazard or undesired event causal factors to the best mitigation selections is proposed. Although there is an emphasis on safety in the report, the same process can also be applied to threat-critical, security-critical, availability-critical, survivability-critical, lethality-critical, or any other mission-critical, or any other mission-critical system or even a combination thereof, in the emerging Network Centric Warfare environment.

The report provides the three basic components of the manual process, as the following,

1. Functional hazard analysis that leads the analyst to identify the hardware, software and human system integration causal factors of each hazard and link each causal factor to one or more mitigating safety-critical function or requirement. In this way the analyst is more able to determine the optimal mitigation solution.

2. Generic hazard analysis that leads to the construction of generic software safety requirements. Generic software safety requirements are those design features, development process "best practices", coding standards and techniques, and other

general requirements that are levied on a system containing safety-critical software, over a range of applications.

3. Traceability of all requirements tagged as safety-critical to test or proof of implementation. Each trace is at the appropriate level of rigor determined by standards documents used by the system designer. Although software is the primary focus of this research effort, the process elements should be applied to hardware and human system integration components as well.

3.5 Product Stories

1. Business Week Design School Article

There was an interesting article in the October 9, 2006 issue of Business Week, entitled "D-Schools". The article discussed that MBA School theory reviews the development of a product focusing on financial analysis and market size. The new Design School theory reviews the development of a product focusing on the customer, allowing for a product to provide different features.

The Design School curriculum includes engineering, business, design and social sciences courses, allowing one person to have a marketing, engineering, designer, and strategist background. A number of companies have multidisciplinary teams, while the Design School provides for a single person to have all the disciplinary capabilities. This approach is similar to Systems Engineering by analyzing all the elements involved in the development a product.

2. Business Week Jack Welsh Article

In the same Business Week magazine there was another interesting article entitled, "Whose Company is it Anyway?" by Jack Welsh (Former CEO of General Electric Corporation) The question asked is "Who is a Company For?"

The answer options are, The Shareholders, The Company Managers, The Company Workers, or The Company Customers. Jack Welsh answer is The Shareholders. I thought this was a logical answer from Jack Welsh, because he took a General Electric Company that was known for building televisions, refrigerators, washers, etc. and made it into a company that moved partly into the banking business, rental of campers and aircraft, etc. He did this because these changes made more money for the Shareholders.

My first question is, what does "Who is a Company For?" mean? Who makes the decisions on a new product developed, including the cost, schedule and quality of the product? In my opinion, this would be the Company Managers. Are their decisions dependent on the Shareholders profit or the Customers wants? A Systems Engineering approach will take all the factors into consideration.

In the Department of Defense, a similar question is asked. "What is your product for or who should the design and development engineers try to satisfy?" The answer options are The Fleet, or the Naval Air Systems Command (NAVAIR), that provided you the funding. A Systems Engineering approach will take all the factors into consideration, including the Fleet, NAVAIR and the design & development managers and engineers.

3. From an article in the Washington Post entitled "The FBI's Upgrade That Wasn't", dated August 18, 2006 the following information was presented. In 2003, Science Applications International Corporation (SAIC) had spent months writing 730,000 lines of computer code for a networked system for tracking criminal cases that was designed to replace the bureau's antiquated paper files. Within a few days the FBI Director Robert S. Mueller III announced that the $170 million system was in serious trouble and a year later the program was cancelled. The nation's premier law enforcement and counterterrorism agency, burdened with one of the government's most archaic computer systems, would have to start from scratch. The collapse of the attempt to remake the FBI's filing system stemmed from failures of almost every kind, including poor conception and muddled execution of the steps needed to make the system work. Lawmakers and experts have faulted the FBI for part of the failed project, because of their poorly written Systems Requirements Document (SRD).

4. From a book entitled "The Commanders" by Bob Woodward, the following Gulf War information was presented.

The Gulf War lasted 42 days. The three air phases took 38 days. The ground war took four days before President Bush declared a cease-fire. The United States and coalition forces overran Kuwait and southern Iraq, destroyed Saddam's army, routed the Republican Guard, dictated the terms of peace and Kuwait was liberated. This situation showed the power of the United States air power.

Chapter 4 System Engineering Process

4.1 Introduction

Systems Engineering transforms an operational need into a final product. The Systems Engineering Life Cycle Process has changed throughout the years. A generic Program Life Cycle Process is presented below, in Figure 4.1.1

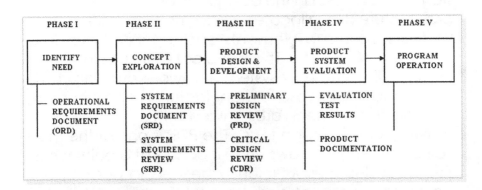

Figure 4.1.1 Program Life Cycle Process

The Operational Requirements Document (ORD) and System Requirements Document (SRD) were detailed in Section 3.2 and 3.3 of this book.

The Evaluation Test Results Document is described in the next Chapter of this book.

The program process includes the System Requirements Review (SRR), Preliminary Design Review (PDR) and the Critical Design Review (CDR), which are described below.

System Requirements Review (SRR): The SRR focuses on the functional and performance requirements of the end product being designed and developed. A key objective of the review is to clear up any uncertainties that may remain from the initial requirements with respect to the specific planned design ands specification requirements. Such a review is most valuable following a competitive source selection, where full and open discussions with the contractors may have been limited prior to contract award.

Preliminary Design Review (PDR): The PDR is held shortly after the detail design of the system sub-elements has started. It focuses on the adequacy of the overall system architecture, the allocated performance and functions for the system sub-elements and the initial design of those sub-elements. This is typically at a point where about 15% of the new system hardware drawings have been ready for release to manufacturing and where the software requirements are finalized with about 10% of the software in unit design tests. The PDR is one of the more valuable design reviews since it occurs at a point where the initial design is sufficiency defined to permit good comprehension, yet early enough in the process that major miscues can be corrected with minimum impact. Proper staffing of the review, with independent oversight and direction, is of paramount importance.

Critical Design Review (CDR): The CDR is held late in the design phase, typically at a point where 85% of the hardware drawings have been ready for release to manufacturing and roughly 85% of software units are ready for software integration. It generally represents the last chance to identify and correct design shortfalls prior to completion of the fabrication and assembly of the product. Serious shortfalls found at this point in the process can have major repercussions with the program costs and schedule. Proper staffing of the review, with independent direction is mandatory.

Each of the Program Phases should address all of the Systems Engineering elements, including the hardware, software, modeling, interface, human factors, security, logistics and safety. In general, the current process in the government utilizes a basic software process that needs to be expanded to include all of the other Systems Engineering elements. The current process, in the government, utilizes The Capability Maturity Model (CMM), developed by Carnegie Mellon University, in 1995. The detailed Capability Maturity Model Software Process is explained in a book entitled "The Capability Maturity Model, Guidelines for Improving the Software Process" by Carnegie Mellon University Software Engineering Institute, 2003. The CMM is a framework that describes the key elements of an effective software process. The CMM describes an evolutionary improvement path for software organization from an ad hoc, immature process to a mature, disciplined one. This path is encompassed by five levels of maturity. The general characteristics and structure of these levels are presented below.

Level 1: Initial

    (1)   Ad Hoc Processes

Level 2: Repeatable

    (1)   Requirements Management
    (2)   Software Project Planning
    (3)   Software Project Tracking & Oversight
    (4)   Software Subcontract Management
    (5)   Software Quality Assurance

    (6)   Software Configuration Management

Level 3: Defined

    (1)   Organization Process Focus
    (2)   Organization Process Definition
    (3)   Training Program
    (4)   Integrated Software Management
    (5)   Software Product Engineering
    (6)   Inter-Group Coordination
    (7)   Peer Review

Level 4: Managed

    (1)   Quantitative Process Management
    (2)   Software Quality Management

Level 5: Optimizing

    (1)   Defect Prevention
    (2)   Technology Change Management
    (3)   Process Change Management

The Capability Maturity Model Process needs to be modified to add all of the Systems Engineering elements into account, including the hardware, software, modeling, interface, human factors, security, logistics and safety. The current Capability Maturity Model process also does not include a detailed performance evaluation process.

4.2 Program Process

4.2.1 Capability Maturity Model Levels 2 and 3

The government is currently utilizing The CMM Level 2 and 3 for the majority of their software development processes. These levels can best be understood with the list of questions that are provided below in Table 4.2.1.1 This Table was developed by Q-Labs Inc. Each CMM Process area identifies Goals, considered important for enhancing the process capability. In addition each CMM Process provides the following features.

1. Commitment: The actions the organization must take to ensure that the process is established and will endure.

2. Ability: Preconditions that must exist to implement the software process.

3. Activity: The roles and procedures necessary to implement the process areas.

4. Measurement: Measure the process areas and analyze the measurement results.

5. Verification: Ensure that the activities performed are in compliance with the process.

TABLE 4.2.1.1 Capability Maturity Model (CMM) Level 2 & 3 Process Questions

REQUIREMENT MANAGEMENT (RM):

RM	**GOAL 1:** System requirements allocated to software are controlled to establish a baseline for software engineering and management use.
RM	The software engineering group reviews the allocated requirements before they are incorporated into the software project.
RM	**GOAL 2:** Software plans, products, and activities are kept consistent with the system requirements allocated to software.
RM	The software engineering group uses the allocated requirements as the basis for software plans, work products, and activities.
RM	Changes to the allocated requirements are reviewed and incorporated into the software project.
RM	**Commitment 1** - The project follows a written organizational policy for managing the system requirements allocated to software.
RM	**Ability 1** - For each project, responsibility is established for analyzing the system requirements and allocating them to hardware, software, and other system components.
RM	**Ability 2** - The allocated requirements are documented.
RM	**Ability 3** - Adequate resources and funding are provided for managing the allocated requirements.
RM	**Ability 4** - Members of the software engineering group and other software-related groups are trained to perform their requirements management activities.
RM	**Measurement 1** - Measurements are made and used to determine the status of the activities for managing the allocated requirements.
RM	**Verification 1** - The activities for managing the allocated requirements are reviewed with senior management on a periodic basis.
RM	**Verification 2** - The activities for managing the allocated requirements are reviewed with the project manager on both a periodic and event-driven basis.
RM	**Verification 3** - The software quality assurance group reviews and/or audits the activities and work products for managing the allocated requirements and reports the results.

(Continued)

	SOFTWARE PROJECT PLANNING (SPP):
SPP	**GOAL 1:** Software estimates are documented for use in planning and tracking software.
SPP	Estimates for the size of the software work products (or changes to the size of software work products) are derived according to a documented procedure.
SPP	Estimates for the software project's effort and costs are derived according to a documented procedure.
SPP	Estimates for the project's critical computer resources are derived according to a documented procedure.
SPP	The project's software schedule is derived according to a documented procedure.
SPP	Software planning data are recorded.
SPP	**GOAL 2:** Software project activities and commitments are planned and documented.
SPP	Software project planning is initiated in the early stages of, and in parallel with, the overall project planning.
SPP	A software life cycle with predefined stages of manageable size is identified or defined.
SPP	The project's software development plan is developed according to a documented procedure.
SPP	The plan for the software project is documented.
SPP	Software work products that are needed to establish and maintain control of the software project are identified.
SPP	The software risks associated with the cost, resource, schedule, and technical aspects of the project are identified, assessed, and documented.
SPP	Plan for the project's software engineering facilities and support tools are prepared.
SPP	**GOAL 3:** Affected groups and individuals agree to their commitments related to the software project.
SPP	The software engineering group participates on the project proposal team.
SPP	The software engineering group participates with other affected groups in the overall project planning throughout the project's life.

(Continued)

SPP	Software project commitments made to individuals and groups external to the organization are reviewed with senior management according to a documented procedure.
SPP	**Commitment 1** - A project software manager is designated to be responsible for negotiating commitments and developing the project's software development plan.
SPP	**Commitment 2** - The project follows a written organizational policy for planning a software project.
SPP	**Ability 1** - A documented and approved statement of work exists for the software project.
SPP	**Ability 2** - Responsibilities for developing the software development plan are assigned.
SPP	**Ability 3** - Adequate resources and funding are provided for planning the software project.
SPP	**Ability 4** - The software managers, software engineers, and other individuals involved in the software project planning are trained in the software estimating and planning procedures applicable to their areas of responsibility.
SPP	**Measurement 1** - Measurements are made and used to determine the status of the software planning activities.
SPP	**Verification 1** - The activities for software project planning are reviewed with senior management on a periodic basis.
SPP	**Verification 2** - The activities for software project planning are reviewed with the project manager on both a periodic and event-driven basis.
SPP	**Verification 3** - The software quality assurance group reviews and/or audits the activities and work products for software project planning and report the results.
SOFTWARE PROJECT TRACKING AND OVERSIGHT (SPTO):	
SPTO	**GOAL 1:** A documented software development plan is used for tracking the software activities and communicating status.
SPTO	The size of the software work products (or size of changes to the software work products) is tracked, and corrective actions are taken as necessary.
SPTO	The project's software effort and costs are tracked, and corrective actions are taken as necessary.
SPTO	The project's critical computer resources are tracked and corrective actions are taken as necessary.

(Continued)

SPTO	The project's software schedule is tracked, and corrective actions are taken as necessary.
SPTO	Software engineering technical activities are tracked, and corrective actions are taken as necessary.
SPTO	The software risks associated with cost, resource, schedule, and technical aspects of the project are tracked.
SPTO	Actual measurement data and replanning data for the software project are recorded.
SPTO	The software engineering groups conduct periodic internal reviews to track technical progress, plans, performance, and issues against the software development plan.
SPTO	Formal reviews to address the accomplishments and results of the software project are conducted at selected project milestones according to a documented procedure.
SPTO	**GOAL 2:** Corrective actions are taken and managed to closure when actual results and performance deviate significantly from the software plans.
SPTO	The project's software development plan is revised according to a documented procedure.
SPTO	**GOAL 3:** Changes to software commitments are agreed to by the affected groups and individuals.
SPTO	Software project commitments and changes to commitments made to individuals and groups external to the organization are reviewed with senior management according to a documented procedure.
SPTO	Approved changes to commitments that affect the software project are communicated to the members of the software engineering group and other software-related groups.
SPTO	**Commitment 1** - A project software manager is designated to be responsible for the project's software activities and results.
SPTO	**Commitment 2** - The project follows a written organizational policy for managing the software project.
SPTO	**Ability 1** - A software development plan for the software project is documented and approved.
SPTO	**Ability 2** - The project software manager explicitly assigns responsibility for software work products and activities.
SPTO	**Ability 3** - Adequate resources and funding are provided for tracking the software project.
SPTO	**Ability 4** - The software managers are trained in managing the technical and personnel aspects of the software project.

(Continued)

SPTO	**Ability 5** - First-line software managers receive orientation in the technical aspects of the software project.
SPTO	**Measurement 1** - Measurements are made and used to determine the status of the software tracking and oversight activities.
SPTO	**Verification 1** - The activities for software project tracking and oversight are reviewed with senior management on a periodic basis.
SPTO	**Verification 2** - The activities for software project tracking and oversight are reviewed with the project manager on both a periodic and event-driven basis.
SPTO	**Verification 3** - The software quality assurance group reviews and/or audits the activities and work products for software project tracking and oversight and report the results.
SOFTWARE SUB-CONTRACT MANAGEMENT (SSM):	
SSM	**GOAL 1:** The prime contractor selects qualified software subcontractors.
SSM	The work to be subcontracted is defined and planned according to a documented procedure.
SSM	The software subcontractor is selected, based on an evaluation of the subcontractor bidders' ability to perform the work, according to a documented procedure.
SSM	**GOAL 2:** The prime contractor and the software subcontractor agree to their commitments to each other.
SSM	The contractual agreement between the prime contractor and the software subcontractor is used as the basis for managing the subcontract.
SSM	A documented subcontractor's software development plan is reviewed and approved by the prime contractor.
SSM	Changes to the software subcontractor's statement of work, subcontract terms and conditions, and other commitments are resolved according to a documented procedure.
SSM	**GOAL 3:** The prime contractor and the software subcontractor maintain ongoing communications.
SSM	The prime contractor's management conducts periodic status/ coordination reviews with the software subcontractor's management.
SSM	Periodic technical reviews and interchanges are held with the software subcontractor.

(Continued)

SSM	Formal reviews to address the subcontractor's software engineering accomplishments and results are conducted at selected milestones according to a documented procedure.
SSM	**GOAL 4:** The prime contractor tracks the software subcontractor actual results and performance against commitments.
SSM	A documented and approved subcontractor's software development plan is used for tracking the software activities and communicating status.
SSM	The prime contractor's software quality assurance group monitors the subcontractor's software quality assurance activities according to a documented procedure.
SSM	The prime contractor's software configuration management group monitors the subcontractor's activities for software configuration management according to a documented procedure.
SSM	The prime contractor conducts acceptance testing as part of the delivery of the subcontractor's software products according to a documented procedure.
SSM	The software subcontractor's performance is evaluated on a periodic basis, and the evaluation is reviewed with the subcontractor.
SSM	**Commitment 1** - The project follows a written organizational policy for managing the software subcontract.
SSM	**Commitment 2** - A subcontract manager is designated to be responsible for establishing and managing the software subcontract.
SSM	**Ability 1** - Adequate resources and funding are provided for selecting the software subcontractor and managing the subcontract.
SSM	**Ability 2** - Software managers and other individuals who are involved in establishing and managing the software subcontract are trained to perform these activities.
SSM	**Ability 3** - Software managers and other individuals who are involved in managing the software subcontract receive orientation in the technical aspects of the subcontract.
SSM	**Measurement 1** - Measurements are made and used to determine the status of the activities for managing the software subcontract.
SSM	**Verification 1** - The activities for managing the software subcontract are reviewed with senior management on a periodic basis.
SSM	**Verification 2** - The activities for managing the software subcontract are reviewed with the project manager on both a periodic and event-driven basis.
SSM	**Verification 3** - The software quality assurance group reviews and/ or audits the activities and work products for managing the software subcontract and report the results.

(Continued)

SOFTWARE QUALITY ASSURANCE (SQA):	
SQA	**GOAL 1:** Software quality assurance activities are planned.
SQA	An SQA plan is prepared for the software project according to a documented procedure.
SQA	**GOAL 2:** Adherence of S/W products and activities to applicable standards, procedures, and requirements is verified objectively.
SQA	The SQA group's activities are performed in accordance with the SQA plan.
SQA	The SQA group participates in the preparation and review of the project's software development plan, standards, and procedures.
SQA	The SQA group reviews the software engineering activities to verify compliance.
SQA	The SQA group audits designated software work products to verify compliance.
SQA	**GOAL 3:** Affected groups and individuals are informed of software quality assurance activities and results.
SQA	The SQA group periodically reports the results of its activities to the software engineering group.
SQA	The SQA group conducts periodic reviews of its activities and findings with the customer's SQA personnel, as appropriate.
SQA	**GOAL 4:** Noncompliance issues that can not be resolved within the software project are addressed by senior management.
SQA	Deviations identified in the software activities and software work products are documented and handled according to a documented procedure.
SQA	**Commitment 1** - The project follows a written organizational policy for implementing software quality assurance (SQA).
SQA	**Ability 1** - A group that is responsible for coordinating and implementing SQA for the project (i.e., the SQA group) exists.
SQA	**Ability 2** - Adequate resources and funding are provided for performing the SQA activities.
SQA	**Ability 3** - Members of the SQA group are trained to perform their SQA activities.
SQA	**Ability 4** - The members of the software project receive orientation on the role, responsibilities, authority, and value of the SQA group.

(Continued)

SQA	**Measurement 1** - Measurements are made and used to determine the cost and schedule status of the SQA activities.
SQA	**Verification 1** - The SQA activities are reviewed with senior management on a periodic basis.
SQA	**Verification 2** - The SQA activities are reviewed with the project manager on both a periodic and event-driven basis.
SQA	**Verification 3** - Experts independent of the SQA group periodically review the activities and software work products of the project's SQA group.

SOFTWARE CONFIGURATION MANAGEMENT (SCM):

SCM	**GOAL 1:** Software configuration management activities are planned.
SCM	An SCM plan is prepared for each software project according to a documented procedure.
SCM	**GOAL 2:** Selected software work products are identified, controlled and available.
SCM	A documented and approved SCM plan is used as the basis for performing the SCM activities.
SCM	A configuration management library system is established as a repository for the software baselines.
SCM	The software work products to be placed under configuration management are identified.
SCM	Products from the software baseline library are created and their release is controlled according to a documented procedure.
SCM	**GOAL 3:** Changes to identified software work products are controlled.
SCM	Change requests and problem reports for all configuration items/units are initiated, recorded, reviewed, approved, and tracked according to a documented procedure.
SCM	Changes to baselines are controlled according to a documented procedure.
SCM	**GOAL 4:** Affected groups and individuals are informed of the status and content of software baselines.
SCM	The status of configuration items/units is recorded according to a documented procedure.
SCM	Standard reports documenting the SCM activities and the contents of the software baseline are developed and made available to affected groups and individuals.

(Continued)

SCM	Software baseline audits are conducted according to a documented procedure.
SCM	**Commitment 1** - The project follows a written organizational policy for implementing software configuration management (SCM).
SCM	**Ability 1** - A board having the authority for managing the project's software baselines (i.e., a software configuration control board-SCCB) exists or is established.
SCM	**Ability 2 -** A group that is responsible for coordinating and implementing SCM for the project (i.e., the SCM group) exists.
SCM	**Ability 3 -** Adequate resources and funding are provided for performing the SCM activities.
SCM	**Ability 4 -** Members of the SCM group are trained in the objectives, procedures, and methods for performing their SCM activities.
SCM	**Ability 5 -** Members of the software engineering group and other software-related groups are trained to perform their SCM activities.
SCM	**Measurement 1** - Measurements are made and used to determine the status of the SCM activities.
SCM	**Verification 1 -** The SCM activities are reviewed with senior management on a periodic basis.
SCM	**Verification 2 -** The SCM activities are reviewed with the project manager on both a periodic and event-driven basis.
SCM	**Verification 3 -** The SCM group periodically audits software baselines to verify that they conform to the documentation that defines them.
SCM	**Verification 4 -** The software quality assurance group reviews and/or audits the activities and work products for SCM and reports the results.
ORGANIZATION PROCESS FOCUS (OPF):	
OPF	**GOAL 1:** Software process development and improvement activities are coordinated across the organization.
OPF	The organization's and projects' activities for developing and improving their software processes are coordinated at the organization level.
OPF	The use of the organization's software process database is coordinated at the organizational level.

(Continued)

OPF	New processes, methods, and tools in limited use in the organization are monitored, evaluated, and, where appropriate, transferred to other parts of the organization.
OPF	Training for the organization's and projects' software processes is coordinated across the organization.
OPF	The groups involved in implementing the software processes are informed of the organizations and projects' activities for software process development and improvement.
OPF	**GOAL 2:** The strengths and weaknesses of the software processes used are identified relative to a process standard.
OPF	The software process is assessed periodically, and action plans are developed to address the assessment findings.
OPF	**GOAL 3:** Organization-level process development and improvement activities are planned.
OPF	The organization develops and maintains a plan for its software process development and improvement activities.
OPF	**Commitment 1** - The organization follows a written organizational policy for coordinating software process development and improvement activities across the organization.
OPF	**Commitment 2** - Senior management sponsors the organization's activities for software process development and improvement.
OPF	**Commitment 3** - Senior management oversees the organization's activities for software process development and improvement.
OPF	**Ability 1** - A group that is responsible for the organization's software process activities exists.
OPF	**Ability 2** - Adequate resources and funding are provided for the organization's software process activities.
OPF	**Ability 3** - Members of the group responsible for the organization's software process activities receive required training to perform these activities.
OPF	**Ability 4** - Members of the software engineering group and other software-related groups receive orientation on the organization's software process activities and their roles in those activities.
OPF	**Measurement 1** - Measurements are made and used to determine the status of the organization's process development and improvement activities.
OPF	**Verification 1** - The activities for software process development and improvement are reviewed with senior management on a periodic basis.

(Continued)

ORGANIZATION PROCESS DEFINITION (OPD):	
OPD	**GOAL 1:** A standard process for the organization is developed and maintained.
OPD	1. The organization's standard software process is developed and maintained according to a documented procedure.
OPD	2. The organization's standard software process is documented according to established organization standards.
OPD	3. Descriptions of software life cycles that are approved for use by the projects are documented and maintained.
OPD	4. Guidelines and criteria for the projects' tailoring of the organization's standard software process are developed and maintained.
OPD	**GOAL 2:** Information related to the use of the organization's standard software process by the software projects is collected, reviewed and made available.
OPD	5. The organization's software process database is established and maintained.
OPD	6. A library of software process-related documentation is established and maintained.
OPD	**Commitment 1** - The organization follows a written policy for developing and maintaining a standard software process and related process assets.
OPD	**Ability 1** - Adequate resources and funding are provided for developing and maintaining the organization's standard software process and related process assets.
OPD	**Ability 2** - The individuals who develop and maintain the organization's standard software process and related process assets receive required training to perform these activities.
OPD	**Measurement 1** - Measurements are made and used to determine the status of the organization's process definition activities.
OPD	**Verification 1** - The software quality assurance group reviews and/or audits the organization's activities and work products for developing and maintaining the organization's standard software process and related process assets and reports the results.
TRAINING PROGRAM (TP):	
TP	**GOAL 1:** Training activities are planned.
TP	Each software project develops and maintains a training plan that specifies its training needs.

(Continued)

TP	The organization's training plan is developed and revised according to a documented procedure.
TP	**GOAL 2:** Training for developing the skills and knowledge needed to perform software management and technical roles is provided.
TP	The training for the organization is performed in accordance with the organization's training plan.
TP	Training courses prepared at the organization level are developed and maintained according to organization standards.
TP	**GOAL 3:** Individuals in the software engineering group and software-related groups receive the training necessary to perform their roles.
TP	A waiver procedure for required training is established and used to determine whether individuals already possess the knowledge and skills required to perform in their designated roles.
TP	Records of training are maintained.
TP	**Commitment 1** - The organization follows a written policy for meeting its training needs.
TP	**Ability 1** - A group responsible for fulfilling the training needs of the organization exists.
TP	**Ability 2** - Adequate resources and funding are provided for implementing the training program.
TP	**Ability 3** - Members of the training group have the necessary skills and knowledge to perform their training activities.
TP	**Ability 4** - Software managers receive orientation on the training program.
TP	**Measurement 1** - Measurements are made and used to determine the status of the training program activities.
TP	**Measurement 2** - Measurements are made and used to determine the quality of the training program.
TP	**Verification 1** - The training program activities are reviewed with senior management on a periodic basis.
TP	**Verification 2** - The training program is independently evaluated on a periodic basis for consistency with, and relevance to, the organization's needs.
TP	**Verification 3** - The training program activities and work products are reviewed and/or audited and the results are reported.

(Continued)

INTEGRATED SOFTWARE MANAGEMENT (ISM):	
ISM	**GOAL 1:** The project's defined software process is a tailored version of the organization's standard process.
ISM	The project's defined software process is developed by tailoring the organization's standard software process according to a documented procedure.
ISM	Each project's defined software process is revised according to a documented procedure.
ISM	The project's software development plan, which describes the use of the project's defined software process, is developed and revised according to a documented procedure.
ISM	**GOAL 2:** The project is planned and manned according to the defined software process.
ISM	The software project is managed in accordance with the project's defined software process.
ISM	The organization's software process database is used for software planning and estimating.
ISM	The size of the software work products (or size of changes to the software work products) is managed according to a documented procedure.
ISM	The project's software effort and costs are managed according to a documented procedure.
ISM	The project's critical computer resources are managed according to a documented procedure.
ISM	The critical dependencies and critical paths of the project's software schedule are managed according to a documented procedure.
ISM	The project's software risks are identified, assessed, documented, and managed according to a documented procedure.
ISM	Reviews of the s/w project are periodically performed to determine the actions needed to bring the s/w project's performance and results in line with the current & projected needs of the business, customer, & end users, as appropriate.
ISM	**Commitment 1** - The project follows a written organizational policy requiring that the software project be planned and managed using the organization's standard software process and related process assets.
ISM	**Ability 1** - Adequate resources and funding are provided for managing the software project using the project's defined software process.

(Continued)

ISM	**Ability 2** - The individuals responsible for developing the project's defined software process receive required training in how to tailor the organization's standard software process and use the related process assets.
ISM	**Ability 3** - The software managers receive required training in managing the technical, administrative, and personnel aspects of the software project based on the project's defined software process.
ISM	**Measurement 1** - Measurements are made and used to determine the effectiveness of the integrated software management activities.
ISM	**Verification 1** - The activities for managing the software project are reviewed with senior management on a periodic basis.
ISM	**Verification 2** - The activities for managing the software project are reviewed with the project manager on both a periodic and event-driven basis.
ISM	**Verification 3** - The software quality assurance group reviews and/or audits the activities and work products for managing the software project and report the results.
SOFTWARE PRODUCT ENGINEERING (SPE):	
SPE	**GOAL 1:** The software engineering tasks are defined, integrated, and consistently performed to produce the software.
SPE	Appropriate software engineering methods and tools are integrated into the project's defined software process.
SPE	The software requirements are developed, maintained, documented, and verified by systematically analyzing the allocated requirements according to the project's defined software process.
SPE	The software design is developed, maintained, documented, and verified, according to the project's defined software process, to accommodate the software requirements and form the framework for coding.
SPE	The software code is developed, maintained, documented, and verified, according to the project's defined software process, to implement the software requirements and software design.
SPE	Software testing is performed according to the project's defined software process.
SPE	Integration testing of the software is planned and performed according to the project's defined software process.
SPE	System and acceptance testing of the software are planned and performed to demonstrate that the software satisfies its requirements.

(Continued)

SPE	The documentation that will be used to operate and maintain the software is developed and maintained according to the project's defined software process.
SPE	Data on defects identified in peer reviews and testing are collected and analyzed according to the project's defined software process.
SPE	**GOAL 2:** Software work products are kept consistent with each other.
SPE	Consistency is maintained across software work products, including the software plans, process descriptions, allocated requirements, software requirements, software design, code, test plans, and test procedures.
SPE	**Commitment 1** - The project follows a written organizational policy for performing the software engineering activities.
SPE	**Ability 1** - Adequate resources and funding are provided for performing the software engineering tasks.
SPE	**Ability 2** - Members of the software engineering technical staff receive required training to perform their technical assignments.
SPE	**Ability 3** - Members of the software engineering technical staff receive orientation in related software engineering disciplines.
SPE	**Ability 4** - The project manager and all software managers receive orientation in the technical aspects of the software project.
SPE	**Measurement 1** - Measurements are made and used to determine the functionality and quality of the software products.
SPE	**Measurement 2** - Measurements are made and used to determine the status of the software product engineering activities.
SPE	**Verification 1** - The activities for software product engineering are reviewed with senior management on a periodic basis.
SPE	**Verification 2** - The activities for software product engineering are reviewed with the project manager on both a periodic and event-driven basis.
SPE	**Verification 3** - The software quality assurance group reviews and/ or audits the activities and work products for software product engineering and reports the results.
INTER-GROUP COORDINATION (IGC):	
IGC	**GOAL 1:** The customer's requirements are agreed to by all affected groups.
IGC	The software engineering group, and the other engineering groups, participate with the customer and end users, as appropriate, to establish the system requirements.

(Continued)

IGC	**GOAL 2:** The commitments between the engineering groups are agreed to by the affected groups.
IGC	A documented plan is used to communicate intergroup commitments and to coordinate and track the work performed.
IGC	Critical dependencies between engineering groups are identified, negotiated, and tracked according to a documented procedure.
IGC	Work products produced as input to other engineering groups are reviewed by representatives of the receiving groups to ensure that the work products meet their needs.
IGC	**GOAL 3:** The engineering groups identify, track, and resolve intergroup issues.
IGC	Representatives of the project's software engineering group work with representatives of the other engineering groups to monitor and coordinate technical activities and resolve technical issues.
IGC	Intergroup issues not resolvable by the individual representatives of the project engineering groups are handled according to a documented procedure.
IGC	Representatives of the project engineering groups conduct periodic technical reviews and interchanges.
IGC	**Commitment 1** - The project follows a written organizational policy for establishing interdisciplinary engineering teams.
IGC	**Ability 1** - Adequate resources and funding are provided for coordinating the software engineering activities with other engineering groups.
IGC	**Ability 2** - The support tools used by the different engineering groups are compatible to enable effective communication and coordination.
IGC	**Ability 3** - All managers in the organization receive required training in teamwork.
IGC	**Ability 4** - All task leaders in each engineering group receive orientation in the processes, methods, and standards used by the other engineering groups.
IGC	**Ability 5** - The members of the engineering groups receive orientation in working as a team.
IGC	**Measurement 1** - Measurements are made and used to determine the status of the intergroup coordination activities.
IGC	**Verification 1** - The activities for intergroup coordination are reviewed with senior management on a periodic basis.

(Continued)

IGC	**Verification 2** - The activities for intergroup coordination are reviewed with the project manager on both a periodic and event-driven basis.
IGC	**Verification 3** - The software quality assurance group reviews and/or audits the activities and work products for intergroup coordination and reports the results.
PEER REVIEWS (PR):	
PR	**GOAL 1:** Peer review activities are planned.
PR	Peer reviews are planned, and the plans are documented.
PR	**GOAL 2:** Defects in the software work products are identified and removed.
PR	Peer reviews are performed according to a documented procedure.
PR	Data on the conduct and results of the peer reviews are recorded.
PR	**Commitment 1** - The project follows a written organizational policy for performing peer reviews.
PR	**Ability 1** - Adequate resources and funding are provided for performing peer reviews on each software work product to be reviewed.
PR	**Ability 2** - Peer review leaders receive required training in how to lead peer reviews.
PR	**Ability 3** - Reviewers who participate in peer reviews receive required training in the objectives, principles, and methods of peer reviews.
PR	**Measurement 1** - Measurements are made and used to determine the status of the peer review activities.
PR	**Verification 1** - The software quality assurance group: reviews and/or audits the activities and work products for peer reviews and reports the results.

In general, the CMM software development process has proven to be an excellent process for producing a good software product, however there are a couple of deficiencies with the CMM process. The first deficiency includes the problem of not using a complete Systems Engineering Approach, including the coordination of the hardware, software, modeling, interface, human factors, security, logistics and safety, to the software process and therefore; some of the Systems Engineering components may be ignored during the execution of the software development process. The second deficiency is that the product performance evaluation process is ignored for both the software and the overall system.

This book will correct the two CMM deficiencies to provide the proper program process.

Chapter 5 System Engineering Performance

5.1 Introduction

Systems Engineering transforms an operational need into a final product. After the product is designed and developed the product must be evaluated to determine if the product meets the required specifications.

During the product design and development life cycle procedure the product should first be completely described in the System Requirements Document (SRD), including the final product performance. The program performance evaluation should include measurement of all of the Systems Engineering elements, including the hardware, software, modeling, interface, human factors, security, logistics and safety. The performance requirements for each of these elements need to be specified in the Systems Requirements Document (SRD). These performance requirements must be identified with specific parameters and each parameter must be identified with a specific number. Ambiguous performance statements will provide you with a product that probably will not work the way you wish.

The SRD should include Operational Requirements (Words) and Performance Specifications (Numbers). The Operational Requirements and Performance Specifications details can be used to measure if the final product meets the expected performance.

Finally, the operational and performance evaluation test plans, such as the list presented below, need to be generated to specifically measure the performance.

- System Test Plan
- System Test Procedure
- Acceptance Test Report
- Validation Test Report

The evaluation test procedure usually includes the following phases

Phase 1: Evaluate the individual components
Phase 2: Evaluate the components in a system situation
Phase 3: Evaluate system rules

Presented below, in Table 5.1.1, is the percentage of time that is spent in the system testing of a typical product.

Table 5.1.1 Program Percent of Effort

Program Phase	Percent Of Effort
Requirement Analysis	6
Preliminary Design	8
Detailed Design	16
Implementation	40
System Testing	20
Acceptance Testing	10

Typically, 30 percent of a program design and development is spent in testing of the product. Testing identifies Discrepancy Reports (DR) of issues that do not match the requirements. Discrepancy Reports quite often number in the thousands, for a complex program, and each of these Discrepancy Report issues must be corrected.

The Capability Maturity Model (CMM) Level 2 and 3 limitations include not specifically pursuing the evaluation of the generated software as well as not providing a complete operational requirements evaluation or a performance specification evaluation, of the final product. Independent Verification and Validation is an absolute requirement in the generation of a system product.

5.2 Program Performance Process

Presented below, in Figure 5.2.1 is a typical program design and development process, which presents the contract tasks needed to evaluate the performance of the program product.

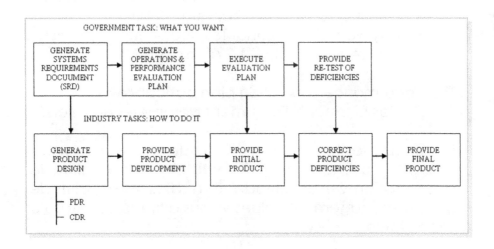

Figure 5.2.1 Program Performance Evaluation Process

The government generates a Systems Requirements Document (SRD), which is usually utilized in the contract Statement of Work. While the contractor produces the program product the government generates several performance evaluation documents that will be used to evaluate the product developed. The performance evaluations usually include an operational evaluation plan and a performance evaluation plan.

The operational evaluation plan basically evaluates the System Requirements Document (SRD) words. For example, for an aircraft model product, the operational evaluation includes determining that the aircraft flies in an X, Y, Z, roll, pitch and yaw direction. The operational evaluation includes determining if the aircraft radar can detect a target and fire a missile at the target. The operational evaluation includes determining if when pressing an aircraft cockpit button the correct result happens. The operational evaluation document, for an aircraft trainer flight simulator, entitled "Trainer Test Performance and Results Report (TTPRR)" is hundreds of pages long and requires weeks to execute. A typical TTPRR page is presented in Chapter 14, Section 14.2, of this book. A typical Operational Test Procedure is provided in Chapter 14, Section 14.3.1.2 of this book.

The performance evaluation plan basically evaluates the System Requirements Document parameter numbers.

For example, for an aircraft product, the real world aircraft maximum altitude, velocity versus altitude, acceleration versus altitude, turn rates versus altitude, energy management values versus altitude, and rate of climb values are evaluated.

Validation measures the real world parameter values against the product parameter values. Verification measures the design parameter values against the product parameter values. The value difference tolerance determines if the product is high fidelity, medium fidelity or low fidelity. The Validation Test Procedures and Results Report provide a very valuable document that provides the overall product performance.

As an example, a Trainer Flight Simulator tactical environment performance evaluation specification is presented below in Table 5.2.1. In Chapter 14, the Trainer Flight Simulator tactical environment performance specifications are presented again, and a complete evaluation of the F-18 Trainer tactical environment is provided.

Table 5.2.1 Trainer Tactical Environment Performance Specification

AIRCRAFT SIMULATION PERFORMANCE

Aircraft Performance Characteristics	Aircraft Simulation Fidelity (% Deviation from Real World Values)		
	High	Medium	Low
(1) Turn Rate Vs Altitude	+ / - 5%	+ / - 10%	+ / - 25%
(2) Energy Management (Turn Rate Vs Altitude)	+ / - 5%	+ / - 10%	+ / - 25%
(3) Climb Rate Vs Altitude	+ / - 5%	+ / - 10%	+ / - 25%
(4) Flight Envelope (Max / Min Speed Vs Altitude)	+ / - 5%	+ / - 10%	+ / - 25%
(5) Acceleration	+ / - 5%	+ / - 10%	+ / - 25%
(6) Database Values	Verification of values from reliable source		

(Continued)

WEAPON SIMULATION PERFORMANCE

Missile Performance Characteristics	Missile Fidelity (% Deviation from Real World Values)		
	High	Medium	Low
(1) Intercept Velocity	+ / - 5%	+ / - 10%	+ / - 25%
(2) Time of Flight	+ / - 5%	+ / - 10%	+ / - 25%
(3) Max / Min Range	+ / - 5%	+ / - 10%	+ / - 25%
(4) Failure Reason	Equivalence	Equivalence	Equivalence
(5) Database Values	Verification of values from reliable source		

RADAR SIMULATION PERFORMANCE

Radar Performance Characteristics	Radar Simulation Fidelity (% Deviation from Real World Values)		
Nose to Nose & Nose to Tail	High	Medium	Low
(1) Detection Range	+ / - 5%	+ / - 10%	+ / - 25%
(2) Acquisition Range	+ / - 5%	+ / - 10%	+ / - 25%
(3) Track Range	+ / - 5%	+ / - 10%	+ / - 25%
(4) Fire Range	+ / - 5%	+ / - 10%	+ / - 25%
(5) Database Values	Verification of values from reliable source		

The operational and performance evaluations are exercised after the product is initially completed. As the operational and performance evaluation tests are conducted, a Discrepancy Report (DR) is generated for each product problem identified. A typical Discrepancy Report (DR) form is presented below in Figure 5.2.3.

(BLACK INK ONLY)

DISCREPANCY REPORT NO:_____ RESPONSIBLE ENGINEER:_____

DEVICE:	S/W BASELINE NO:_____	DATE:_____
TEST:_____ BOOK:_____	PG:_____	TTPRR REV:_____
TEST NAME:_____		CAT/CODE:_____
ORIGINATOR:_____		TEC REPORT:_____

DEFICIENCY:

CORRECTIVE ACTION:

RECHECK:	PASS	FAIL	DATE	NAME
#1	____	____	____	_____
#2	____	____	____	_____

COMMENTS:

DEFICIENCY CLOSED:

NAWC-TSD:_____ DATE:_____

(NAWC-TSD SIGNATURE MUST BE PRESENT FOR CLOSURE)

Figure 5.2.3 Discrepancy Report Form

The Discrepancy Reports are provided with a priority of importance. The contractor corrects the problem detailed in the Discrepancy Report, usually in the order of the most important priority. The trainer flight simulators that I have helped evaluate quite often provided more than one thousand Discrepancy Reports, during the program evaluation cycle.

The evaluation cycle, including the correction of the majority of the Discrepancy Reports, quite often required a couple of years of the program development cycle. The list of program Discrepancy Reports (DR) must be carefully documented with a Configuration Management system.

Chapter 6 Program Success Factors

6.1 Introduction

A program requires a number of factors to be positive in order for the final product to be successful. Some of the factors involved include the following,

- Organization Factor
- Management Factor
- People Factor
- Base Realignment and Closure (BRAC) Factor
- Promotion Factor
- Contracts Factor
- Travel Factor
- Presentation Factor
- Ethics Factor
- Sexual Harassment Factor
- Personal Problems Factor
- Change Factor

What do these factors have to do with Systems Engineering? A person once said that a Software Engineer is interested in paying attention to the software viewgraphs, in a presentation, but a Systems Engineer is interested in paying attention to all the viewgraphs, in a presentation. Each of the factors listed above are an important program element that need to be addressed in order that a program product development is successful. Each one of these factors will be discussed in this chapter of the book.

6.2 Organization Factor

6.2.1 Organization Type

Throughout the years there have been many organization structures that have been used to provide for the development of a successful product. These organization structures, and variations of these structures, include organizations that provide the leadership to the Functional Group Leaders (Department Heads, Division Heads, and Branch Heads) and organizations that provide the leadership to the Program Manager.

The current Defense Department Organization operation (Implemented in 1995) allows the Program Manager to have almost total control of the product development decisions. The Functional Group Leaders responsibility is to train their personnel, analyze how to improve the product development process, and provide the personnel to a specific program, upon request from a Program Manager. The Functional Group Leader does provide a strong input in the decision to develop a product using in-house personnel versus contractor personnel. This organization provides the engineer more individual responsibility for their product accomplishments.

The previous Defense Department Organization operation (Prior to 1995) provided for the Functional Group Leaders to have control of the product development decisions. The Program Manager had to follow the directions of the Functional Group Leader. The main difficulty with this organization was that the Functional Group Leader quite often catered to the organization problems instead of the program problems. A program person quite often had to work on an organization effort using program funding. Another major problem I found was that a product document, such as a simple trip report, had to be reviewed by the Functional Group Leader before it could be distributed. This quite often did not allow a trip report to be distributed for a month or more after the trip.

In my opinion, the new Defense Department Organization operation is far superior to the previous operation.

There are many organization types. Some of the Specific Functional Group Organizations include the following,

The Systems Engineering Organization Diagram, presented in Figure 6.3.1.1, provides all of the elements of System Engineering. The Navy's Training Organization in Orlando Florida uses an organization structure very close to this structure.

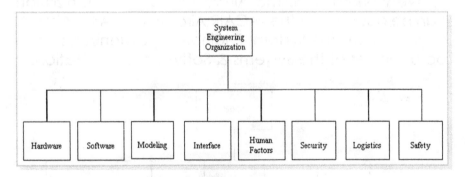

Figure 6.2.1.1 Systems Engineering Organization Diagram

The Aircraft Product Organization, presented in Figure 6.2.1.2, provides a number of Navy Aircraft that will be worked on with a variety of design and development efforts.

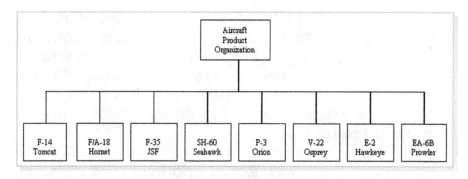

Figure 6.2.1.2 Aircraft Product Organization Diagram

The Matrix Organization, presented in Figure 6.2.1.3, provides for a number of high levels engineering areas, including general aircraft products, such as TACAIR Aircraft, ASW Aircraft, and Navigation, Crew Station and Avionics and a number of general engineering disciplines, such as Analysis, Systems & Software and Safety. The Matrix Organization is the type of organization that has been utilized by most of the Navy Bases that I have worked with. The Aircraft Product Organization can be a sub-set of the TACAIR Aircraft and ASW Aircraft Organization. The Systems Engineering Organization can be a sub-set of the Systems & Software Organization.

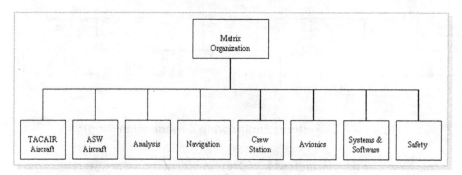

Figure 6.2.1.3 Matrix Organization Diagram

The Organization Factor can be a Systems Engineering problem that can affect the success of providing a good product.

6.2.2 Organization Stories

I have not been directly involved in leading an organization, in my 48 years of Defense Department service, because I have tried to concentrate on the technical side of the design and development of a product. However, I have always belonged to an organization and have seen many of the difficulties that organization leaders have had to endure. I also have been acting Branch Head, Acting Division Head, for a week, from time to time. I have been a Project Coordinator Leader for a couple of years, which was actually a technical Division Head. I have been a Program Manager on a number of technical programs and always micro-managed to the point that I did as much technical work on the program as any of the other engineers on the program.

Presented below are a couple of stories of my involvement with new organizations throughout my career,

1. Naval Air Warfare Center, Patuxent River, Software Division (1996-2006)

In FY2000, the Software Division consisted of about 100 persons, including a Division Head, four Branch Heads, one assistant Branch Head, three secretaries, two financial analysts and about 90 technical persons. The technical persons worked on various engineering programs and were paid by these programs. The staff was paid with overhead funding. Each Division Head, at the Patuxent River Naval Base, was provided overhead funding each fiscal year which paid for their workers on overhead, engineering training, special efforts, yearly awards, etc. I personally knew very little about the distribution of the division funding, but in 2000 the Division Head provided a presentation to everyone in the Software Division organization. He said that his Software Division had a deficit for the year of $600,000. The Division Head was force to reduce his overhead costs, by reducing his staff to two Branch Heads, and during the next couple of years eliminating all division secretaries, and all financial analysts. This story details some of the difficulties of being a Division Head in an organization that is trying to reduce the Defense Department budget.

2. Base Realignment and Closure (BRAC) Organization (1996)

The new Patuxent River Naval Air Warfare Center (NAWC) Organization, Department Heads, Division Heads and Branch Heads were determined through a series of interviews and the results were that all of the Naval Air Systems Command (NAVAIR) workers won all of the Department Head and Division Head positions. All of the Warminster Naval Air Development Center managers were demoted to the Branch Head level. Persons that were a Division Head for 25 years were demoted to a Branch Head position. The new managers were, in general, very good, but it was sad to see a number of my friends demoted.

3. Organization Leader Change (1975-1985)

In the 1960's and 1970's I worked for a Systems Engineering Branch Head that introduced me to the Systems Engineering concept. In 1975 this Branch Head was sent to the War College, in Rhode Island, for a year of education. The Branch Head was not gone for more than a couple of days when we were told that our Systems Engineering Branch was being taken over by a Software Branch Head. The new Branch Head did not know any of us so we had to start over with any promotional possibilities. I personally was on the verge of a promotion that was then postponed for another six years.

I always seem to work for managers that lost an organization battle, so I indirectly also lost the same battle.

The Software Branch Head was later promoted to be the head of a new Flight Simulation Division that was created, with our old Systems Engineering Branch hardware engineers and his Software Branch software engineers.

He selected his software engineers to be the new Branch Heads.

The Flight Simulator Division remained in place for about five years, but was then taken over by a new Department Head. The Flight Simulator Division Head was demoted to a Branch Head and we received a new Fight Simulator Division Head. The former Flight Simulator Division Head said that this was just about the worst thing that ever happened to him in his life, but of course he obtained the organization in 1975, in a take over maneuver.

The new Flight Simulator Division Head opened the promotion possibility to everyone in his division and after a very long promotion procedure, in 1981; I was awarded the GS-13 position. There was 500 GS-12 positions at the Naval Air Development Center, at this time, and about 20 promotions a years were being awarded.

The Flight Simulator Division Head later became the Department Head and "believe it our not" our former Flight Simulator Branch Head once again became the new Flight Simulator Division Head. I talked to him the day before the position selection and he said that he didn't think he had a chance to win back the position. Winning an organization position depends strongly on the committee that selects the person for the position.

6.3 Management Factor

A good manager will enhance the product development, with professional direction, in a courteous, tactful manner. A manager should be responsive to workers requests, provide fairness to all workers, be honest and open and if the worker performs well, reward them. A manager will also set a good example, through their work ethic and personal actions with their workers. A manager will provide a shared responsibility relationship, provides a bi-directional flow of information, provides a free flow of ideas and again, treat all personnel as professionals.

The most important task of a current Functional Group Leader is to provide specific training in a variety of engineering disciplines that will make an engineer more valuable to the organizations programs.

A manager's most important job is to motivate his workers. The first step is to find the person that can do the job and then find a way to keep them happy. Motivating people on your team, figuring out what makes them work hard, is not always an easy task. The work world has some workers with bad attitudes; workers with the ability to do the job, but who are just not motivated to do the job. These are the workers who make the managers job a real challenge.

Listening to what a worker has to say about the program or issue is important and helps to show that you are interested in their ideas. Rewarding a worker for doing a good job is the best way to motivate a worker and the reward quite often can be simply verbal praise.

Managing different workers requires different rules. Some workers can be given a task with some instructions and a period later they will provide you with a product, without any further guidance. Some workers require daily guidance. A manager must be flexible enough to use different techniques to manage different workers.

One of the difficult situations for a manager is what to do with a worker that just is not capable of providing a reasonable product for a program. Jack Welsh, from General Electric, fired 10 % of his workers each year, but the government does not easily allow workers to be fired. In the government, a Program Manager usually tells the Functional Group Manager that they are out of funding for a specific poor worker and the Functional Group Manager moves the poor worker to another organization during a re-organization period. Sometimes a poor worker can be put into a situation that fits their capability and the poor worker becomes a good worker.

From an article entitled "How to deal with Difficult People" by Robert M. Bramson, Ph D, there are a number of worker personalities that might make a program success difficult, including the following,

1. Dictators: These people need every detail done their way. They attack your work, criticize your work, and if they don't get their way, quite often they have a tantrum over your work. They are the Sherman Tanks that come out charging with an "attack" demeanor. They are abusive, intimidating and overwhelming. The Snipers maintain a cover and takes pot shots at you. The Exploders presents a fearsome attack filled with rage that barely seems under control.

Countermeasures to this personality include the following,

- Standing up to them without fighting
- Give the issue time to go away
- Seek group approval regarding the issue

2. Bulldozers: They are highly productive people, through and accurate thinkers who make competent, careful plans and then carry them through, even when the obstacles are great. So how can they be difficult? These people are sometimes beyond moral doubt, hard to dissuade even when their plan appears to be headed for failure and if it does fail they blame others as incompetent.

Countermeasures to this personality include the following,

- Provide alternative views while carefully avoiding direct challenges to their expertise.
- Try to get them to change, if you feel their direction is bad, by possibly getting group approval.

3. Know it Alls: These people speak with great authority about subjects they have little knowledge. Quite often they spread bewilderment wherever they go. The Know it All, unlike the Bull Dozer, has a calm manner and tone and will almost beg you to believe them.

Countermeasures to this personality include the following,

- Present your data as an alternative set of facts rather than the only set of facts.
- Provide the "Know it All" a face saving way out when they are wrong

4. Yes People: They manipulate the presentation of reality in order to gain your approval. They make commitments that they cannot or will not follow through on. They have a tendency to tell you what they think you want to hear.

Countermeasures to this personality include the following,

- Work hard to surface the facts that prevent the Yes-People from taking action.
- Ask non-threatening questions that will provide honest answers, but let them know that you value their work and opinions.
- Don't allow them to make unrealistic commitments.

5. No People: They are negative and pessimistic and will constantly point out why something will not work. They are inflexible and resist change. Their attitude will cause problems with the entire organization.

Countermeasures to this personality include the following,

- State your own realistic optimism trying to connect your statements with past successes in solving similar problems.
- Try to persuade them out of their pessimism, pointing out alternative plans.

6. Passive People: Avoiding conflict and controversy at all costs, these people never offer ideas or opinions, and never let you know where they stand. No matter what you say to them they only respond with a "Yes" or "No"

Countermeasures to this personality include the following,

- Let your tone imply that you expect an answer. Be friendly and encouraging. If they still will not talk explain that you believe that it is very important to talk about the issues and that you expect a discussion next time you meet.
- If the Passive Person opens up, listen carefully and don't talk too much.

7. Complainers: Nothing is ever right with these people. They would rather complain about things than change them. They have a negative affect on the entire program team.

Countermeasures to this personality include the following,

- Listen attentively their complaints and then try to move to a problem solving phase by suggesting alternative ways to elevate the problem.
- Try to move to a problem-solving mode.

The Management Factor can be a Systems Engineering problem that can affect the success of providing a good product.

6.3.1 Management Factor Stories

Presented below are a couple of stories of my involvement with managers throughout my career,

1. Management to Management Disagreement on Program Responsibility (1982 to 1985)

In the early 1980's I helped design and develop the Dynamic Flight Simulator. The Dynamic Flight Simulator simulates the F-14A Aircraft, which was incorporated into the Naval Air Development Center (NADC) Centrifuge, to conduct aircraft spin simulations. The NADC Centrifuge, illustrated below in Figure 6.3.1.1, is a 10 ft diameter sphere mounted on a 50 ft long arm. The NADC Centrifuge produced up to 40 G of force in the X, Y, and Z axis directions. The Dynamic Flight Simulator provides for an allowable force of 15 G's. An aircraft spin usually produces about 7 or 8 G's force. Many years ago monkeys were utilized to study G's forces in the NADC Centrifuge. Also, the Mercury and Gemini Astronauts were trained in the NADC Centrifuge.

Figure 6.3.1.1 Naval Air Development Center Centrifuge

The Dynamic Flight Simulator (DFS) was designed to have an F-14A Aircraft Crewstation that provides 15 G's in all axis directions. The "out-the-window" visual display system, providing the real world view was designed to survive 15 G's in all axis directions. The flight simulation computer that provided the flight simulator instrument, flight controls, switch closure, and light signals was also designed to survive 15 G's in all axis directions. A single 1553 Bus transported the various signals through one of the centrifuge slip rings and then the computer distributed the signals within the centrifuge. The F-14A Aircraft flight controls used just two centrifuge hydraulic inputs.

The Dynamic Flight Simulator exercised pilots in a spin simulation environment and generated an aircraft display that identifies to the pilot the clock wise or counter clock wise direction of the spin and whether the aircraft is right side up or up-side down. The pilot applied the proper controls to eliminate the spin.

As a Systems Engineer, I basically designed the Dynamic Flight Simulator. The Dynamic Flight Simulator was managed by the Crew Systems Department. My Software Department Head, hearing about the design and the fact that our Dynamic Flight Simulator Team was moving full speed to build the system asked the question, "Where is the Dynamic Flight Simulator Requirements Document?" We did not have requirement document. My Department Head provided the funding and a supplement to our team to generate a Dynamic Flight Simulator Design Document. The document took many months to generate and I ended up generating more than half of the document.

I totally agree that we should have had a design document, and the generation of the document provided some insights in our design and development direction. The problem was that now that the Software Department Head provided the Dynamic Flight Simulator requirements they wanted to take over the responsibility of providing the design and development of the system. Also, since I had basically designed the Dynamic Flight Simulator, and worked for the Software Department, this would add to the validity of taking over the program. The centrifuge ownership however belonged to the Crew System Department, so they thought that they should retain the responsibility of the program. I don't know what happened during the responsibility discussions with the Technical Director of our base, but the Crew Systems Department retained the responsibility of the program.

Pressure was put on me to drop off of the Dynamic Flight Simulator Program, but since I basically designed the system and the Technical Director was enthusiastic about the program, I stayed on the program until it was completed.

When President Bush, as Vice President, visited the Naval Air Development Center he toured the Dynamic Flight Simulator. My Software Department management gave me poor performance grades while I worked on the program, because I did not drop off of the program when they pressured me to quit.

A technical worker quite often has to accept the decisions or disagreements between managers and still attempt to provide a quality product. Unfortunately, it is difficult to separate decisions made that you have very little input with a program that you are totally involved with.

---

2. Management to Management Disagreement on Lab
   Takeover (1990's)

In the early 1990's I helped designed the TACAIR Systems
Development Facility (TSDF). The Flight Simulation Division,
which is part of the Systems and Software Department,
traded their ownership of the Analog Computer area
for funding to build the TACAIR Systems Development
Facility.

The initial TACAIR Systems Development Facility area
was 80 ft long by 20 ft high by 20 ft high. I took the
$75,000 in funding provided and designed a three section
facility, which included two high bay Flight Simulator
areas (30 ft by 20 ft by 20 ft high), on either side of a
low bay Problem Control Station area (20 ft by 20 ft
by 8 ft high). The plan was to provide a TACAIR Flight
Simulator in one bay and an Anti-Submarine Warfare
(ASW) Flight Simulator in the other bay, but the ASW
Flight Simulator was not sold before we moved to the
Patuxent River Naval Base. The Problem Control Station
was designed to enable a number of mission exercises,
including incorporation of bombs on the F-14 Aircraft and
studying new Head-Up presentations.

An F-14D Aircraft Flight Simulator, illustrated below
in Figure 6.3.1.2, was designed and developed that
included a forward and rear cockpit crewstation, with
a full set of instrumentation, displays, flight controls,
panels, switches and indicators. The Flight Simulator
included an "out-the-window" visual display system
that provided a real world image over a 60 degree
horizontal by 40 degree vertical viewing area. The
Research Flight Simulator includes a complete cockpit
versus computer interface system.

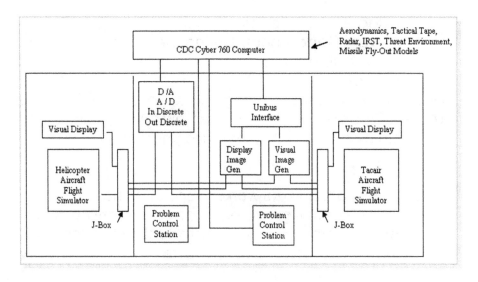

Figure 6.3.1.2 Tacair System Development Facility (TSDF)

The TACAIR System Development Facility (TSDF) design and development required about two years to fully complete, but once completed the facility was determined to be a valuable asset. The funding for the design and development was obtained from the F-14 Aircraft Class Desk in the Naval Air System Command, in Crystal City Virginia.

During a demonstration of the F-14D Aircraft missile delivery exercise, to a large number of Foreign Admirals, including Admirals from Russia, China, Israel, etc. my engineer could not hit the target aircraft. After a number of tries, the Russian Admiral said I can fly the F-14D Aircraft, fire the missile and hit the target. I politely said that I did not think that I was allowed to let him fly the aircraft for security reasons.

The TACAIR Department thought that the TACAIR System Development Facility (TSDF) was a valuable asset and decided to take it over as their facility from the System and Software Department. During the meeting I pointed out, in a vocal manner, that this facility was designed and developed by me, but the System and Software Department was silent during the takeover meeting. The TACAIR Department had a past history of taking over valuable assets that had been designed and developed by others.

Once again, a technical worker quite often has to accept the decisions or disagreements between managers and still attempt to provide a quality product. Unfortunately, it is difficult to separate decisions made that you have very little input with a program that you are totally involved with.

3. Worker versus Manager Disagreement on Lab Emergency Parts Purchase (1990's)

I have helped design and develop many Aircraft Flight Simulator Labs, including the F-14A Aircraft, F-14D Aircraft, LAMPS Anti-Submarine SH-2F / SH-60 Aircraft, CH-53 Aircraft, and F-111 Aircraft. For each Flight Simulator a number of experimental design exercises were conducted. Quite often a number of fleet crewpersons were involved in the experimental design exercises and quite often a part was needed, in an emergency, to ensure that the flight simulator was operating properly. The procurement system at our Naval Base, did not allow you to obtain a part in an emergency, so quite often the fleet crewpersons were standing around with nothing to do.

On one occasion the TSDF F-14D Flight Simulator was to be demonstrated to the Naval Air Systems Command manager of labs, and a five volt power supply burned out, not allowing the "out-the-window" visual display system to operate. I called the visual display system company to send out a five volt power supply replacement, on an emergency basis, and put in the procurement paperwork, with funding, to the Supply Department. When the five volt power supply came in, the following day, without all the proper procurement paperwork being processed, I was called into the Supply Department Main Office and was threatened with being fired. Our Naval Base bureaucracy did not allow program problems to be easily solved.

A company drove up from Florida with an avionic system that I purchased, but arrived too late on Friday (at 3 pm) for the Supply Department to properly process the paperwork. The company was told that they would have to return the following Monday. I put the system in my lab and told the company driver that I would have the paperwork processed on Monday, so that the driver could return to Florida that night. On Monday, I went over to the Supply Department with the paperwork, and was told that I had to return to the Supply Department Main Office, for another reprimand. I thought I might be Jean Val Jean.

There were many situations in which I drove over to a store, such as Radio Shack, and bought parts with my own money, so that I could get a flight simulator operational during an experimental design exercise.

This was not true at all Naval Bases. During one exercise at the Norfolk Naval Base, a strip chart recorder would not operate. The engineer in charge called a store and a new strip recorder was delivered in less than an hour. He said the procurement paper work would follow.

I generated many memorandums requesting an emergency procurement system. Several other engineers also requested an emergency procurement system and after many years one was put into operation. A credit card system was put into place to allow an engineer to procure a part in an emergency situation, but the system does not allow the part to be obtained as fast as the strip chart recorder was obtained at the Norfolk Naval Facility.

An engineer must work within the rules of the organization, but sometimes the rules do not allow the program to be completed on schedule.

4. Manager Absent Mindedness (1980's)

I worked for a good, somewhat famous, but somewhat absent minded manager on the Dynamic Flight Simulator, in the 1980's. The Dynamic Flight Simulator was an F-14A Flight Simulator, which used the Naval Air Development Center Centrifuge System, to conduct aircraft spin exercises. The program needed a million dollars in funding to purchase an "out-the-window" visual display system. Arrangements were made for the Dynamic Flight Simulator Team to make a presentation to the Director of Navy Labs, in Washington D. C., to request the visual display system funding. I took my family to Washington D.C. a day early so that after the presentation we could tour the city.

I arrived at the Director of Navy Labs Office an hour early for the presentation. A short time later the Department Head and Division Head showed up at the Director of Navy Labs Office. The time for the presentation arrived and my Dynamic Flight Simulator Manager had not arrived. The Department Head and Division Head were getting nervous. About 15 minutes late, our Program Manager arrived, somewhat disheveled. The presentation went well, the funding was provided, and the program was successful, so "Alls well that ends well.", but the rest of the story must be told.

The Dynamic Flight Simulator Program Manager drove to the Philadelphia Airport to fly to Washington D.C. He checked his briefcase when he arrived at the Washington D.C Airport, to see if he had all the viewgraphs, he needed for his presentation and discovered that he had forgotten to put the viewgraphs in his briefcase. The Manager was also not sure where he had left the set of viewgraphs; at his office desk, at home or elsewhere. He called his local contractor to check his desk at the office. He called his son to check for the viewgraphs at his home. The viewgraphs were not found, but the local contractor made a new set of viewgraphs, drove them to the Philadelphia Airport, bought a seat for the viewgraphs, and flew them to the Washington D.C. Airport. The Manager went into the aircraft, picked up the viewgraphs, hailed a taxi cab, and arrived at the Director of Navy Labs Office, just 15 minutes late. As far as I know, the original viewgraphs have never been found.

I guess I should have carried a spare set of viewgraphs to protect an absent minded, but otherwise good and dedicated manager.

6.4 People Factor

My friend Ted Hungerford once said to me, "You give me the right person that can produce a product and the program will be a complete success." I totally agree with this statement. The success of any product development depends on the managers and engineers that are working on the program. The difficulty is usually in putting together all of the best managers and engineers that can ensure the program success, and then keeping the people together as a team.

As a worker, professionalism can be defined as a worker that is knowledgeable in their craft, provide a full days work for a days pay and conduct their work with integrity.

My advice to my children was to, "Always do your Best". If all the persons on your program do their best I'm sure your program will be a success.

In general, I think Federal Civil Service Workers sometimes get an unfair reputation. The same worker that is criticized as a federal worker is quite often hired as an industry workers and suddenly become a great worker. They don't suddenly become a good worker, they are good workers.

6.4.1 Intelligence, Maturity, Motivation and Interest Factor

In my opinion, the four main elements of a person's ability to accomplish an effort are their intelligence, maturity, motivation and interest. Intelligence can be measured, but maturity, motivation and interest are somewhat subjective.

When I interviewed a person for a position, motivation and interest were very important to me. It was always interesting to evaluate a person after they worked for a couple of years to see if your initial evaluation was correct.

The government used to have a program in which very intelligent high school persons were given a scholarship to the college of their choice, with the requirement that after they graduated they had to work for the government for at least 1-1/2 years. This seemed like a very good idea, but most of the scholarship persons did not fulfill their 1-1/2 year working obligation and many of them did not pay back their scholarship money. This program has not been used for a number of years.

If a person's manager has the Dictator personality, described in the previous Manager Section of this book my advise is to try to work out the issue with your immediate manager and do not try to solve the issue by going over their head, to the next level of management. Your immediate manager does not like you to talk to their manager about an issue between you and themselves, and will probably get angry with this approach

The People Factor can be a Systems Engineering problem that can affect the success of providing a good product.

6.5 Defense Department Base Realignment and Closure (BRAC) Commission Factor

Defense Authorization Amendments and Base Closure and Realignment Act (P.L. Law No. 100-526, Oct 24, 1988 and U.S.C. Law No. 2687, established the Base Closure process. The Base Closure Commission, appointed by the Secretary of Defense, chose the bases to be closed or realigned. The Secretary of Defense approved the Commission list and forwarded the list to congress. This list became law if Congress did not enact a resolution of disapproval.

The Secretary Of Defense Dick Cheney, in the Base Realignment and Closure (BRAC) Report, said, " In reviewing our need for forces through the mid-1990's and in light of declining defense budgets, we have identified bases in the United States we no longer need and locations oversees where we can end our operations or reduce our forces. It is essential that we reduce the number of installations where a shrinking force is based if we are to get the greatest value from a declining defense budget."

The BRAC base closure selection criteria included the following,

- The current and future mission capability.
- The condition of the land, facilities and air space.
- The ability to accommodate change, mobilization and future force requirements.
- The cost savings of operations and manpower.
- The impact on the existing community.
- The environmental impact.

- The ability of the receiving community to support the missions and personnel.

The first four BRAC Rounds resulted in 97 major base closures, including the following.

1. BRAC 88- 16 total (4 Navy, 5 Air Force, 7 Army)

2. BRAC 91- 26 total (8 Navy, 13 Air Force, 1 Marine, 4 Army)

3. BRAC 93- 28 total (19 Navy, 6 Air Force, 1 Marine, 1 Army)

4. BRAC 95- 27 total (9 Navy, 5 Air Force, 11 Army)

On Sunday, June 30, 1991, the Base Realignment and Closure Commission voted to move the Naval Air Development Center (NADC), in Warminster, Pennsylvania to the Patuxent River Naval Base in Patuxent River, Maryland. The move was to occur by 30 September 1996. 1800 of 2800 NADC workers were offered positions at the Patuxent River Naval Base.

The normal psychological feelings of denial, anger and then acceptance prevailed with many Warminster workers, but after the move was completed, in September 1996, it was determined that the Base Realignment and Closure (BRAC) Commission decision was a fairly good decision. Only 450 NADC workers accepted a position at the Patuxent River Naval Base, but many other NADC workers did retire and join contractors working at the Patuxent River Naval Base. Three major multi-million dollar Research and Development Buildings were constructed for the NADC workers and these buildings proved to be a very valuable asset to the Navy and the Patuxent River Naval Base. Some engineers were hired to fill the positions of NADC workers not moving to Maryland.

In my opinion, the Base Realignment and Closure Commission (BRAC) decision was a good decision, but obviously all of the NADC programs under development were not completed on their current schedule, because plans for the 1996 move took precedence. The closing of bases was a major factor in not providing certain products, on schedule.

The Patuxent River Naval Air Station complex stretches across 25 miles of shoreline at the mouth of the Patuxent River, overlooking the Chesapeake Bay, 65 miles southeast of Washington DC. The complex covers approximately 6,500 acres of airfields, with a long main runway, engineering buildings and support personal housing, at the Navy facility.

The Naval Air Warfare Center (NAWC) today serves as the Navy's principal research, development, test, evaluation, engineering and fleet support activity for naval aircraft, engines, avionics, aircraft support systems and ship/shore/ air operations. The facility also hosts the U. S. Navy Test Pilot School.

Presented below, in Figure 6.5.1, is a Locator Map of the Patuxent River Navy Base, as of the year 2003.

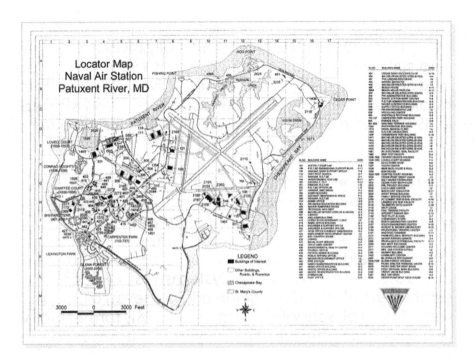

Figure 6.5.1 Patuxent River Locator Map

Presented below is a brief history of the Naval Air Station, Patuxent River obtained from the Global Security.org website.

The Naval Air Station, Patuxent River, was commissioned on 1 April 1943, in an effort to centralize widely dispersed air testing established during the pre-World War II years. Spurred on by events of World War II the consolidation effort was swift. Flight Test operations were provided after ground was broken in 1942. Rear Admiral John S. McCain, then Chief of the Navy's Bureau of Aeronautics, called Patuxent River "the most needed station in the Navy" during the commissioning ceremony on 1 April 1943.

The Naval Air Test Center was established as a separate entity on June 16, 1945, organizationally dividing the test and support functions. During World War II, hundreds of combat experienced pilots arrived here to test airplanes. Formalized classroom instruction started in 1948 with the establishment of a Test Pilot Training Division.

During the Korean War, the Patuxent River Naval Base was faced with developing jet aircraft and improving existing conventional weapons for the war effort. In 1953, the Tactical Test Division merged with the Service Test Division. The U.S. Naval Test Pilot School was established in 1958 and the Weapons Systems Test Division was established in 1960. Four of the original seven astronauts were Test Pilot School graduates. In 1961, former Navy test pilot Alan Shepard became the first American in space.

In 1975, the Naval Air Test Center (NATC) was made the Naval Air Systems Command's principal site for Strike Aircraft, Anti-submarine Aircraft, Rotary Wing Aircraft and Systems Engineering development testing.

On 1 January 1992, the Naval Air Warfare Center (NAWC), Aircraft Division, was identified as the new organization name at Patuxent River.

Over $155 million, in military construction was spent for new engineering complexes and renovation of existing facilities as a result of the Base Realignment and Closure (BRAC) actions that located personnel from Naval Air Warfare Center (NAWC), Aircraft Division sites at Warminster, Pennsylvania and Trenton, New Jersey to Patuxent River. The Naval Air Systems Command, located in Crystal City, Virginia was also relocated in Patuxent

River, in 1997, as a result of the Base Realignment and Closure legislation.

With the consolidation, the Navy has become the largest employer in the Patuxent River community. More than 17,000 people are now stationed at the complex. The Naval Air Station at Patuxent River is now known as the center of excellence for naval aviation. Patuxent River hosts the full spectrum of acquisition management, research and development capabilities, air and ground test and evaluation, aircraft logistics and maintenance management. This distinctive synergy supports land based and maritime aircraft and engineering, test and evaluation, integration and life cycle support for ship/ shore electronics. These combined capabilities are unique within the Department of Defense and ensure the Naval Air Station Patuxent River's status as an aviation leader working effectively to continue progress into the 21st Century.

The Base Realignment and Closure (BRAC) Factor can be a Systems Engineering problem that can affect the success of providing a good product.

6.5.1 Base Realignment and Closure (BRAC) Stories

1. BRAC Organization Decisions:

The Base Realignment and Closure (BRAC) Commission moved our Naval Air Development Center, in Warminster Pennsylvania to the Patuxent River Naval Base, Maryland, in 1996. The following year, in 1997, the BRAC Commission also moved The Naval Air System Command (NAVAIR), from Crystal City, Virginia to the Patuxent River Naval Base. The Naval Air Warfare Center (NAWC) Aircraft Division provides the technical persons for a program. The Naval Air Systems Command (NAVAIR) provides the funding and general management for the program.

The new Naval Air Warfare Center (NAWC) Organization, Department Heads, Division Heads and Branch Heads were determined through a series of interviews and the results were that all of the Naval Air Systems Command (NAVAIR) workers won all of the Department Head and Division Head positions. All of the Naval Air Development Center managers were demoted to the Branch Head level. Persons that were a Division Head for 25 years were demoted to a Branch Head position. The new managers were, in general, very good, but it was sad to see a number of my friends demoted.

In the new Base Realignment and Closure (BRAC) organization, I was made the Project Coordinator for the Naval Air Systems Command (NAVAIR) Aircraft Training Organization (PMA-205). The new organization operation did not allow the Department Head, Division Head and Branch Head to review program details, so the organization needed a technical person to review the program details. About 25 Naval Air Warfare Center (NAWC) Project Coordinators were created to line up with each Naval Air Systems Command (NAVAIR) Organization. This position was maintained for about five years and then was dissolved. The Project Coordinator position was developed by the Naval Air Warfare Center (NAWC) and the Naval Air System Command (NAVAIR) did not see the value of the position.

2. BRAC Move Process

Everyone that made the decision to accept the position at the Patuxent Naval Base had to then decide if they wanted to sell their house in the Warminster area and buy a house in the Patuxent River area. My wife and I decided to sell our house in Holland, Pennsylvania and build a new house in Waldorf, Maryland. This is obviously a major effort for anyone that accepted the new position, but if you are in the military this effort happens every couple of years. Our neighbor, directly across the street in Holland, with a house very similar to ours, sold their house during this period for $235,000, so this is the figure I used for our sale. After nine months on the market and 100 families visiting our house we did not have one offer to sell our house. In the meanwhile we built a new house in Waldorf, Maryland for $235,000 and our transfer date of 15 September 1996 had arrived. I had to re-structure our new house mortgage which ended up costing more money and providing a higher interest rate. Fortunely, our house did sell not long after we moved to Maryland, but the offer was for $190,000, which we accepted, because paying for two house mortgages is not an easy budget situation. The BRAC move had some difficult issues for many of the people involved.

6.6 Promotion Factor

The most important factor that all workers want is a promotion. Basically, the main reason workers have a job is to earn money and obtaining additional money from time to time is an absolute necessary to encourage them to perform at a high level of performance.

For many organizations, the promotion policy is not totally clear and so the worker is confused what they have to do to be rewarded a promotion. For many organizations promotion levels are very limited so obtaining additional money is somewhat difficult.

In the government there are GS Levels of promotions. A person is usually hired at a GS-5 or GS-7 Level. The GS-9, GS-11 and the GS-12 Levels are usually automatically obtained over about a 5 year period. The next GS promotional levels, which are the GS-13, GS-14 and GS-15 Levels, are then obtained over the next 25 to 35 years. The federal government provides all workers a small, usually about 3 % of your salary, cost-of-living raise each year. The federal government also provides a performance award, usually about $500 to $1000, each year that dependent on your manager's subjective evaluation.

I was hired in 1958 at a GS-5 Level. It took me 9 years to obtain my GS-12 Level, 23 years to obtain my GS-13 Level, and 28 years to obtain the GS-14. I then remained at a GS-14 Level for the next 20 years until retirement. I was satisfied with this situation, because my motivation was to stay in the technical area of work, rather than move into the management area of work, which is where most of the high level promotions are obtained in the government.

I was talking to a worker in 2006, who is a good worker, but desperately needed money for college tuition, because he had two children in college at the same time. He had received his GS-14 Level about five years ago and he now wanted his GS-15 Level to obtain additional money for the college bills. The person was angry, very angry at every manager, because he could not obtain a promotion. Throughout the years I have seen many workers that were very angry that they could not get a promotion. No matter how many times I said to this person, "Don't be angry" he remained very angry, and in fact decided to retire from the government, after 30 years of service, to pursue a job that will provide more money for his children's college bills.

A worker obtaining a promotion is quite often a major factor in the success of a program. If your key worker is un-happy with their salary then your program might be in trouble.

In my opinion, the criteria for obtaining a promotion should be detailed in a written document. The criteria can include undergraduate education, graduate education, program performance, patents, years of service, years within grade, and manager's subjective evaluation. A worker should be evaluated every year and the manager should review the evaluation with each worker. Unfortunately, there are only a number of promotions that can be provided each year and the persons that do not obtain the promotion are often un-happy. A very good Division Manager, in our organization, once opened a GS-13 Promotion to every one in his Division. Fifteen people applied, were evaluated, and a couple of persons were selected for the GS-13 promotion. Many of the persons that were not selected immediately left the organization, because they finally realized where they fit on the promotional cycle.

The Promotion Factor can be a Systems Engineering problem that can affect the success of providing a good product.

6.7 Contracts Factor

In the 1950's and 1960''s most of the products accomplished at the Naval Air Development Center (NADC) were done in-house. There were massive machine shops, electrical shops, and integration/installation shops. At some point in time it was decided that all work will be contracted out which totally changed the work of a typical NADC engineer. The new work included the generation of a contract package, attending design reviews to evaluate the contractor work progress, the generation of a product evaluation plan, and the execution of the evaluation test plan.

From the website http://en.wikipedia.org/wiki/federal_acquisition_regulation, the following contract history is presented.

The United States government exercises its contract powers through legislation and regulations through the Constitution. A contract with the United States Government must comply with the laws and regulations that permit it, and made by the Contracting Officer with authority to make the contract. The regulations are provided in the Federal Acquisition regulation (FAR) and the Contracting Officer ensures that these regulations are followed.

Contract types are generally grouped into two broad categories: Fixed Price Contracts and Cost Plus Fixed Fee Contracts.

A Fixed Price Contract includes the following.

- A price that is not subject to any adjustments.
- Places the contractor maximum risk and full responsibility for all costs and resulting profit.

- It provides maximum incentive for the contractor to control costs and perform effectively.
- The Fixed Priced Contract is the preferred method of contracting from the government's perspective.

Variations of Fixed Price Contracts include the following.

- Economic Price Adjustment Contract: Revision of prices for specific contingencies, such as change in labor or material costs.
- Incentive Contracts: Provides profit motive for the contractor to perform efficiently from a cost perspective.

A Cost Plus Fixed Fee Contract includes the following.

- Provides for payment of allowable incurred costs, to the extent prescribed in the contract.
- Cost reimbursement contracts place the least cost and performance risk on the contractor.
- Cost reimbursement contracts are suitable for use only when uncertainties involved in contract performance do not permit costs to be estimated with sufficient accuracy to provide a fixed priced contract.
- Used for research and development contracts where the cost of the effort is not easily determined.

Variations of Cost Plus Fixed Fee Contracts include the following.

- Time and Material Contract: Direct labor hours at specified hourly rates, and materials at cost.
- Letter Contract: Authorize the contractor to begin work immediately manufacturing supplies or performing services.
- Indefinite Delivery Contracts: Provides an indefinite quantity, within stated limits, of supplies or services during a fixed period.
- Agreements: A written instrument of understanding, negotiated between a contracting activity and a contractor, reviewing uncertain requirements, supply quantities, costs, and contemplating future contracts.
- Purchase Orders: Fixed price for acquisition of commercial items.
- Government Commercial Purchase Card: Purchase supplies with credit card.

A contract package must be generated to obtain proposals from industry to produce a successful product.

The contract package includes the following elements,

1. Cover Letter: The cover letter includes the program name, description, and estimated cost.

2. Procurement Checklist: The checklist is the contract procurement elements presented in this list.

3. Statement of Work: The Contract Statement of Work provides all of the operations and performance details

of the product, that is required for a contractor to design, develop and produce the product.

4. Government Cost Estimate: The estimated product cost includes the labor, travel and material level of effort.

5. Source List: A list of all the companies that may provide a proposal to develop the product.

6. Justification Document: The Justification and Approval Document is provided for a contract other than a full and open competition. The Justification Document includes the description of need, statutory authority, the contractors' unique qualifications, market research (Commerce Business Daily) announcement, determination of a fair and reasonable cost, actions taken to remove barriers to competition, and the signature of the Program Manager, Legal Counsel, and Contracting Officer.

7. Evaluation Plan: The plan that will be used to evaluate the final program product.

8. Proprietary List: A list of individuals that had access to proprietary or source selection contract information.

9. Personnel Qualifications: This is a list of the personnel types that are needed to accomplish the development of the product. Some of the personnel types could be the Principal Engineer, Senior Aeronautical Engineer, Aeronautical Engineer, Weapon System Analyst, Software Engineer, Junior Engineer, Test Pilot Analyst, Training System Analyst, Technical Aide, Draftsperson and a Clerical Person.

---

10. Contract Data Requirement's List (DD-1423): This is a list of all documents that are needed to accomplish the development of the product. The Contract Data Requirement's includes the Document Number, Document Title, Document Data Item Description (DID), Statement of Work paragraph, frequency of delivery, and general remarks.

11. Contract Security Classification Specification (DD-254): The Contract Security Classification Specification includes a list of contractor general access requirements, such as restricted data or intelligence data and a list of contractor specific access requirements. such as receive classified documents or material.

12. Government Furnished Property and Information (GFP & GFI) List: This is a list of the hardware, software, and modeling elements that the government will provide during the development of the product.

13. Contracting Officer's Representative (COR) Nomination Letter: The COR Nomination Letter provides for an engineer to be a technical representative of the Contracting Officer. The duties include providing technical direction and guidance to the specifications or the Statement of Work (SOW), under the contract. The COR also periodically checks the performance of the contractor's product.

14. (ADP) Approval Document: This document provides analysis that the computers involved in the contract product have the correct capabilities.

15. Funding Document: This document provides proof that there is funding for the contract product.

6.7.1 Contracting Analysis

Contractor work with the government is based on winning the contract Request for Proposal (RFP), so they do everything they can, within the law, to win the contract. A number of strategies are used to win a contract.

A Sole Source Contract eliminates any competition and all contractors try to influence you to provide a Sole Source Contract. In many cases if you have a contractor that is providing you personnel service and you think they have done an excellent job, you would like the contract to be sole source. The Contracting Officer makes the decision if the Request for Proposal will be competitive or sole source. My experience is the Contracting Officer made all my efforts competitive.

There are also government engineering strategies that can be used. You can ask for a Competitive Negotiated Contract as opposed to a Competitive "Low Bidder Wins" Contract. You can then set up the following evaluation criteria,

1. Technical Approach   50%
2. Personnel   25%
3. Facilities   12.5%
4. Past Work Efforts   12.5%
5. Cost   25%

This criterion allows the engineer to weigh somewhat in favor of the company that they feel could do the best job for their program. A "Low Bidder Wins" Contract quite often allows a company to "buy in" to a program, even though you may feel that they might not provide a successful program.

Another issue or strategy that an engineer can address is whether the contact will be a Level of Effort or Delivery Order Contract. A Level of Effort Contract is usually pursued if the delivery efforts are clearly understood. Once this contract is signed, funding can be put on the contract without any negotiation. A Delivery Order Contract is usually pursued if the delivery efforts are not clearly defined, but each funding increment put on the contract must be negotiated.   Once again the Contracting Officer makes the decision whether the contract will be a Level of Effort Contract or a Delivery Order Contract. My experience is that the Contracting Officer will allow you to have a Level of Effort Contract if you provide them a good reason why you want to go in that direction.

The Contracting Factor can be a Systems Engineering problem that can affect the success of producing a good product.

6.7.2 Contracting Stories

Throughout the years I have helped negotiate about 50 various types of service and material contracts that the Contracting Officer signed and in most cases the service or product provided was very good. Presented below are a couple of stories of contract negotiations that were not very successful.

1. Contracting Strategy (1990's)

In 1990's, I submitted a Request for Proposal for a five year Systems Engineering Support Contract to industry. I was very happy with the contractor that was currently doing the job. The Contracting Officer determined that the Request for Proposal will be competitive. The Contracting Officer determined that the Request for Proposal will be a "Low Bidder Wins" Contract. My current contractor put in a very competitive bid, but the company that was determined to be the low bidder said that their workers would work 48 hours for 40 hours of pay, providing them the lowest bid. I was forced to accept the company's bid, get rid of the contractor persons working on the program, and train the new persons that won the contract. And as far as I could determine, I never got 48 hours of work for 40 hours of pay. The company did however provide a fairly good support service and the program that picked up my previous support persons said they were delighted with their performance.

The lesson learned is to insist on using a Negotiated Contract with evaluation criteria that only has the cost as part of the final contract determination.

2. Small Business Contract (1990's)

The Small Business Office continually tried to convince you put a contract out to a Small Business Company, which could be accomplished with a Sole Source Contract. In the 1990's I allowed a sole source contract to be signed to develop a number of F-14D Aircraft Flight Simulator panels to be provided by a Small Business Contractor, and I thought they did a good job.

I then needed to build a special Liquid Crystal Display and wanted to send out a competitive Request for Proposal. During an industry search I found a company that could produce the special Liquid Crystal Display. The Small Business Office said that because I gave the last display contract to a Small Business Company by law I had to send the Liquid Crystal Display contract to the same company. The Small Business Company then agreed to sub-contract the development of the Liquid Crystal Display to the company that could produce the product. The contract was signed.

The Small Business Company immediately went bankrupted. They must have known they were going to file for bankruptcy when they signed the contract. I demanded that my funding be sent directly to the company that could produce my Liquid Crystal Display and was told by the Small Business Office and the Contracting Officer that legally this was not possible. I was angry, but could do nothing with the situation.

The lesson learned is to be more careful with the type of contact that you allow to be signed.

3. Contract Statement of Work Problem (1990's)

A Statement of Work was generated for the purchase of a television monitor. The statement of work did not include a specific request for an on/off switch, so the contractor provided a system in which you had to turn the system on and off by pulling out the cable from the electrical outlet. The Statement of Work did not state that the television monitor had to have a new CRT system, so the contractor provided a CRT with a couple of burn marks. In general, the equipment purchased was very poor.

The lesson learned was to be very careful when generating the contract Statement of Work. This other lesson learned was to never buy another product from this fairly well known company.

6.8 Travel Factor

During the first 10 years of employment almost no travel occurred. Systems, such as the design and development of a flight simulator, were built in-house, using several in-house machine shops, electrical shops and installation shops. All of the flight simulator software was also developed in-house. Laws were then passed to have industry provide all government systems, such as a Research or Trainer Flight Simulator. The government contracted out all program systems and a series of design reviews were established. Many programs were also pursued as joint Navy, Air force and Army programs. This change obviously required a great deal of travel for program meetings and program design reviews. Communications were not yet developed to have a video meeting between program managers and engineers so that you had to travel to a central location for discussions. One year, I was required to travel 26 times. This was my highest travel total of my career. Throughout my career I have traveled to just about every city in the United States.

The travel factor is an issue that just has to be taken into account in the success of any program. I always found that engineers getting together and talking over issues on a program had a positive affect on the program success. Also, it seemed that the more inputs to the program, the better chance were for a program success. I urge that managers listen to the young engineers input on a program, because quite often they have a positive input.

Presented below, in Table 6.8.1, is a list of travel situations that I have pursued in my 48 years working for the Department of Defense. As seen in Table 6.8.1, travel was continually pursued to the Navy Training Center, in Orlando, Florida and the Navy Training Headquarters, in Washington D. C. The Navy Training Headquarters moved to Paturent River, Maryland, in 1998. My Navy Base, in Warminster, Pennsylvania, moved to the Navy Base in Patuxent River, Maryland, in 1996. The F-14D Trainer was designed and developed by McDonnell Douglas Corporation, in St Louis, Missouri and the AAI Company, in Hunt Valley, Maryland. The F-14B/A Trainer was designed and developed by Grumman Aircraft Company, in Long Island, New York. The F-18 Trainer was designed and developed by L3 Communications Company, in Dallas, Texas. The F-14B/A & D Trainers were located in San Diego, California and Virginia Beach, Virginia. The F-18 Trainers are currently located in Virginia Beach, Virginia and Lemoore, California.

The Travel Factor can be a Systems Engineering problem that can affect the success of providing a good product.

Table 6.8.1 Travel Details		
CITY, STATE	**PLACE**	**NUMBER OF TIMES**
Ann Arbor, Michigan	Industry	2
Albuquerque, New Mexico	Air Force Base	6
Allendale, New Jersey	Industry	2
Arlington, Virginia	Industry	3
Ashville, North Carolina	Industry	5
Binghamton, New York	Industry	4
China Lake, California	Navy Base	3
Dahlgren, Virginia	Navy Base	4
Dallas, Texas	Industry (F-18 Trainer)	4
Dayton, Ohio	Air Force Base	4
Daytona Beach, Florida	Industry	2
Dothan, Alabama	Army Base	1
El Centro, California	Navy Base	4
Elmira, New York	Industry	1
Fallon, Nevada	Navy Base	1
Fort Walton Beach, Florida	Air Force Base	3
Huntsville, Alabama	Navy Base	1
Hunt Valley, Maryland	Industry (F-14 Trainer)	8
Indianapolis, Indiana	Navy Base	2
Jacksonville, Florida	Navy Base	1

(Continued)

CITY, STATE	PLACE	NUMBER OF TIMES
Key West, Florida	Navy Base	3
Long Island, New York	Industry (F-14 Trainer)	10
Madison, Wisconsin	University	1
New Orleans, Louisiana	Navy Base	3
Norfolk, Virginia	Navy Base (SH-2F Trainer)	12
Norwood, New Jersey	Industry	2
Orlando, Florida	Navy Base (Training Center)	30
Pensacola, Florida	Navy Base	2
Pt Mugu, California	Navy Base	10
Phoenix, Arizona	Air Force Base	4
Patuxent River, Maryland	Navy Base	10
San Diego, California	Navy Base (F-14 Trainer)	16
St Louis, Missouri	Industry (F-14 Trainer)	16
San Antonio, Texas	Air Force Base	3
Salt Lake City, Nevada	Industry	4
Stamford, Connecticut	Industry	2
Syracuse, New York	Industry	1
Virginia Beach, Virginia	Navy Base (F-18 & F14 Trainer)	27
Washington, D. C.	Navy Trainer Headquarters	30
Yuma, Arizona	Navy Base	1
Total Number of Travel Times		245

6.8.1 Travel Stories

Many of my program traveling situations was very successful; however there are always bad travel stories. Presented below are a couple of my stories.

1. Motel Accommodations:

In the 1970's I was traveling to El Centro, California for the CH-46 Helicopter In-flight Escape System (HEPS) Program. It was the cricket bug season and the bugs were piled high in the streets. My motel room had a number of crickets in the room. It was evening and I was tired so I just picked them up and put them outside of my motel room and went to sleep. I was awakened by a cricket on my face. I turned the lights on to find dozens of crickets in the room. What could I do? I just pulled the sheets over my heads and tried to sleep the night.

In the 1980's, the government decided that they could save travel money by having government workers stay in a Bachelor Officer Quarters (BOQ) when they traveled. The cost per night was about $10, so this was a large savings in funding. In those days, however, the BOQ's did not have a television or telephone in the room and you had to share a bathroom with a stranger from another room. These arrangements were not comfortable for many years, but eventually the BOQ's were updated with everything provided by a regular motel and the cost was much less than the motel cost.

2. Aircraft Accommodations:

During one flight from Philadelphia to the West Coast the aircraft lost one of its two engines and we had to quickly return to Philadelphia. In Philadelphia, the maintenance workers tried to repair the engine, and the airlines would not give us our luggage to make other arrangements for our trip. I was determined not to fly that same aircraft to California. I found a way to get into the room that my luggage was being stored and would not leave until they allowed me to take my baggage with me.

During one flight from Albuquerque, New Mexico back to the Baltimore Washington Airport, I had to stop in four cities on the way. This was the only arrangements that would guarantee me a seat on the aircraft.

To fly back to San Diego, California from El Centro, California in order to get an aircraft back home I had to fly in an open air two seat crop duster aircraft over the local mountains. I did not like this flight.

3. Travel Arrangements:

In the 1970's, on a trip to Ashville, North Carolina I was made to drive my car to the end of a Philadelphia subway line, and then took a taxi cab to an airport. I then flew to Washington D.C., changed airlines, and flew to Ashville, North Carolina. On the way back I flew to Washington D.C., but then had to take a taxi cab to the train station so I could take a train to Philadelphia. I then took the subway to the end of the line, where my car was parked. I could not convince the Travel Department that this was not a reasonable travel arrangement.

In the 1990's there were riots in Los Angeles due to a court decision dealing with the Rodney King incident. The following week, I needed to travel to the Point Mugu Navy Base, which is 60 miles north of Los Angeles. Usually I would fly to Los Angeles and then rent a car and drive to the Point Mugu Navy Base. I put in a travel request to fly to San Francisco, and then to Santa Barbara and then to drive south to the Point Mugu Navy Base. These travel arrangements were somewhat more costly than the past travel arrangements but I thought they would be much safer for me. The travel arrangements were turned down by the Naval Air Development Center Technical Director. I postponed the Point Mugu Navy Base Meeting.

4. Home Problems during Travel

One year I had to travel 26 times and when I came home on one Friday evening I found that someone had tried to break into our house and had broken all of the dead bolt locks on every door, trying to break in. Someone in our family had lost their key to our house. The dead bolt locks could be put on a "No" situation at night, so that the key would not work from the outside of the house. The potential robber tried to get through the dead bolt locks only to break each lock. I had to replace every dead bolt lock that weekend to allow me to travel on that Monday. I could not quickly buy the "Yes" / "No" deadbolt lock, so I put two normal deadbolt locks on each door and did not allow one key to ever leave the house.   This deadbolt lock was only used at night. I completed the work on the weekend and went on a business trip on Monday.

6.9 Presentation Factor

One of the most important program factors in the success of producing a good product is providing presentations, which helps to obtain funding for the program, provides understanding of the design and development of the product, and allows discussions on the technical details of the program product. It is extremely important that good clear presentations are provided by the organization that is in need of the product, the organization that is providing the design and development of the product, and the organization that is evaluating the product performance.

There is a variety of presentation methods, but the best method appears to be a presentation using a number of viewgraphs and a well spoken person presenting the viewgraphs.

Someone once said that a typical engineer basically listens to the details of the viewgraphs that are specifically in their engineering technical area, but a Systems Engineer listens carefully to the details of all of the viewgraphs.

In my opinion, two factors that are negative factors in a presentation are the use of cliché's and acronyms.

Some cliché's and acronyms in a presentation are alright, but providing many cliché's and acronyms in a technical presentation is not professional and quite often does not allow the audience to understand the presentation details.

Some of the cliché's that I personally do not like to hear in a presentation include the following. "Get a leg up", "get your ducks in a row", "as we speak", "beat a dead horse".

An example of using too many acronyms in a presentation is the following. "The ASW LAMPS Program includes MAD, DIFAR, LOFAR, and CASS systems.

6.10 Ethics Factor

Ethics is defined as a system of moral principles or values. In business, ethics is defined as the rules or standards governing the conduct of the members of the profession. Each business has a different set of ethic rules so it is important that the specific ethic rules for your business are completely understood and that the rules are enforced by the business management. The ethics rules are usually complicated and lengthy so it is very difficult to sometimes know if you should or should not do something. Accepting a meal from someone you are doing business with is an ethic issue that changes from time to time and business to business so you must know what is the current rule for your workplace.

From the website http://www.dod.mil/dodgc/defense_ethics the following Department of Defense ethic rules are presented.

Gifts:

1. A gift may be accepted if it is valued at less than $20, as per C. F. R. Law No. 2635.204. A gift based on a personal relationship, such as a family relationship or personal relationship may also be accepted, at any value.

2. Awards may be accepted, if they are a bona fide award, and if the cash or gift value is less than $200, as per C. F. R. Law No. 2635.204. Award gift in excess of a $200 value require written determination from agency ethics official.

Conflict of Interest:

1. Federal employees may not participate personally and substantially in any official capacity in which they have a personal interest, as per C.F.R. Law No. 2635.402. A review of the issue with the base lawyer is advised.

2. Federal employees may not participate personally and substantially in any official capacity in which they have a financial interest, as per U.S.C. Law No. 208 and C.F.R. Law No. 2635.402. This includes financial holdings such as stocks, bonds, leasehold interests, mineral and property rights, deeds, trust, liens, options or commodity futures. In the Defense Department, a Financial Disclosure must be generated each year and reviewed by the base lawyer to determine if there is any conflict of interest with any program issues.

3. Federal employees may not participate personally and substantially in any official capacity in which they are seeking employment, as per U.S.C. Law No. 423. A review of the issue with the base lawyer is advised. A one year ban applies to persons who participate in the negotiations of a contract in excess of $10 million.

FAR Law No.3.104-1 prohibits, for two years, after leaving federal service a former employee from communicating before any agency of the government with intent to influence a particular matter.

4. Federal employees may not obtain a private gain from a travel arrangement, as per C. F. R. Law No. 2635.101. Passenger carriers may only be used for official purposes. Only persons whose transportation benefits

the government should use government transportation. Government transportation should be arranged to be cost effective. Government transportation should not favor rank or position. Government transportation should use commercial transportation.

The Ethics Factor can be a Systems Engineering problem that can affect the success of producing a good product.

6.10.1 Ethics Stories

1. Travel Abuse:

An engineer in our Department created travel plans for a vacation, pretending that this trip was for a program meeting.

The government paid for these trips. Since the program was a secret program the checks and balances that would expose this fraud, did not occur. Somehow, the problem was finally identified and the engineer fired.

2. Equipment Abuse:

An engineer in the Supply Department took equipment home and sold them in a store he owned near his home.

This situation was also identified and the engineer fired.

3. Job Information Abuse:

A Software Engineer obtained his job with our base by using a college transcript from a college he did not attend.

He simply wrote to a college, using his real name and asked for the school transcript with that name. He actually obtained a couple of college transcripts and used the one with the highest grades. He worked for our Division and was a very good and knowledgeable worker. When he was confronted and fired he said, "How can you fire a person that is doing a good job?"

4. Manager Abuse:

The Ethics Rules were yearly presented, by the base lawyer, but these rules were occasionally broken. A couple of base managers would sign a major contract with a contractor just before they would leave the government and obtain a job with that contractor. One base manager provided many dozens, of sole source contracts to a contractor and he later obtained a position with that same contractor.

6. 11 Sexual Harassment Factor

From the website http://en.wikipedia.org/wiki/sexual_ harassment, the following information is obtained.

Sexual harassment is a form of sex discrimination that violates Title VII of the Civil Rights Act of 1964. Unwelcome sexual advances, requests for sexual favors, and other verbal or physical of a sexual nature constitute Sexual harassment when this conduct explicitly or implicitly affects an individual's employment, unreasonably interferes with an individual's work performance, or creates an intimidating, hostile or offensive work environment.

Sexual harassment can occur in a variety of circumstances, including but not limited to the following:

1. The victim as well as the harasser may be a women or a man. The victim does not have to be of the opposite sex.

2. The harasser can be the victim's supervisor, an agent of the employer, a supervisor in another area, a co-worker, or a non-employee.

3. The victim does not have to be the person harassed but could be anyone affected by the offensive conduct.

4. Unlawful sexual harassment may occur without economic injury to or discharge of the victim.

5. The harasser's conduct must be unwelcome.

It is helpful for the victim to inform the harasser directly that the conduct is unwelcome and must stop. The victim should use any employer complaint mechanism or grievance system available.

When investigating allegations of sexual harassment, the whole record needs to be reviewed, such as the nature of the sexual advances, and the context in which the alleged incidents occurred. A determination on the allegations is made from the facts on a case by case basis.

Prevention is the best tool to eliminate sexual harassment in the workplace. Employers are encouraged to take steps necessary to prevent sexual harassment from occurring. They should clearly communicate to employees that sexual harassment will not be tolerated. They can do so by providing sexual harassment training to their employees and by establishing an effective complaint or grievance process and taking immediate and appropriate action when an employee complains.

It is also unlawful to retaliate against an individual for opposing employment practices that discriminate based on sex or for filing a discrimination charge testifying, or participating in any way in an investigation, proceeding, or litigation under Title VII.

I can't remember a great deal of major sexual harassment incidents that occurred at our Navy Base other than inappropriate jokes, inappropriate comments or inappropriate pictures displayed. Most of these incidents were simply ignored except by the very sensitive persons, but management has attempted to reduce the incidents. The Navy military did have the Tailhook Parties, which were not at all appropriate, and these incidents were hopefully brought under control. Sexual Harassment training with management enforcement has provided progress of personnel being more sensitive to sexual harassment.

Sexual Harassment definitely has a major affect on the successful accomplishment of a program.

From a Business Week Article, dated 13 November 2006, sexual harassment charges filed with the Equal Employment Opportunity Commission fell from 15,549 in 1995 to 12,679 in 2005. The article suggested the definition of harassment has changed from 20 years to today, as presented below.

20 Years Ago	Today
1. Sexually explicit remarks	Sexually explicit e-mails
2. Lewd one-on-one talk	Raunchy office banter
3. Inappropriate touching	Inappropriate joking
4. Violating office romance policies	Violating company relationship packs

The Sexual Harassment Factor can be a Systems Engineering problem that can affect the success of producing a good product.

6.11.1 Sexual Harassment Stories

1. Inappropriate Advances:

A rumor had one of our F-14 Aircraft Trainer Program Managers having sexual relations with a female contractor worker and a fact occurred that the Program Manager was relieved of his position. The sexual harassment process seemed to work very well in this situation.

2. Inappropriate Activity:

It was rumored that one software support worker at the Warminster Naval Base was providing prostitution services in her van at lunch on many occasions. She was fired after she was caught after one lunch rendezvous.

3. Inappropriate Pictures:

A number of years ago inappropriate pictures were displayed throughout the Warminster Naval Base. In the 1980's this practice was declared not allowable, and all of the pictures were removed.

6.12 Personnel Problems Factor

Every organization has personnel with personal problems. If these personal problems are severe then quite often the program success suffers. Sometimes the personal problems are not caused by the person affected, such as a health problem of a spouse or one of their children, but quite often the personal problem is caused by the individual. Many rumors of persons having an affair or some other personal problem are quite often known by the program manager and other program workers and these problems cause some difficulty with the programs success. I will say that some personnel can produce excellent program work even though they are involved in a personal situation, but on the other hand personals problems or affairs usually affect a program in a negative way.

The Personal Problem Factor can be a Systems Engineering problem that can affect the success of producing a good product.

6.12.1 Personnel Problem Stories

1. Persons with Personal Problems

I had an electrical technician worker with severe alcoholic problems, and he rarely showed up on a Monday. When he did show up for work, he did an excellent job, so I basically tolerated his situation.

An acquaintance of mine was having an affair with his neighbor, who had three children, only to later find out that her husband was having an affair with his wife, who also had three children. They simply swapped spouses, which was strange to me.

_____

A friend of mine's wife walked out on him and left him to raise their six children by himself. He is writing a book on his difficulty of raising six children while producing excellent work for our base.

My wife had severe health problems during my primary working years, which required us to confront difficult situations. Also, two of my four children also had health problems during their years of development. I tried to separate my work time with my family time, but this is sometimes difficult.

One person I traveled with went to a bar each night to drink alcohol and even brought a bottle of liquor in his luggage. The alcohol did not seem to affect his work and he was my boss, so I tolerated the situation.

One married person I traveled with went out each night to look for women and when the women he met the previous night did not respond to his advances he wasvery disappointed at the meeting the following day.

2. Program Health Issues

One issue that is not always explained in a technical book is some of the issues that are going on behind the scenes, during a program effort. During the design, development and evaluation of the SH-2F Helicopter Trainer Flight Simulator Update Program (1984-86), a number of health issues occurred. My wife had a pituitary tumor operation in January 1985. I ate at a restaurant in Norfolk, Virginia, in May 1985, and was provided with e-coli, which was not completely understood in 1985. I was very sick for months, but did not miss more than a couple of days of work, over this period. My very knowledgable aerodynamic modeling engineer was working 12 hour days, seven days a week, for about six months and ended up with psychological problems that required him to be treated at a mental hospital for a period of time after the program ended. A Captain from NAVAIR called me aside one day and had a list of complaints about this modeling engineer that one day he took a two hour lunch and he took off one Sunday afternoon. I said, "Are you kidding me?" The main PAX River Evaluation Team Pilot was killed in a SH-2F Helicopter accident, during our program. The streets in NAS Patuxent Naval Base are named after test pilots that have been killed in aircraft accidents.

6.13 Change Factor

Most people do not like change and strongly resist change. In any occupation change happens and personal must adapt to it. The Base Realignment and Closure (BRAC) exercise caused the most traumatic changes for most of my friends at the Warminster Naval Base and you could see the stress of many of my fellow engineers. Of 1800 workers invited to transfer to the Patuxent River Naval Base only 450 workers made the decision to transfer.

Change also includes being transferred to another organization, transferred to another program, or given a different assignment on your current program. If the change is your choice then the change is not stressful, but if the change is not your choice then the change can be very stressful.

What seems like a normal organization change to a manager quite often is a very stressful change to a worker

The Change Factor can be a Systems Engineering problem that can affect the success of producing a good product. A manager can anticipate the difficulty of a change and provide counseling regarding the change.

Several engineers became very depressed during the BRAC Process.

Two fellow engineers were put into a new organization one day and that night they had a heart attack, and one of them passed away.

6.14 Technology Transfer Factor

The DoD Technology Transfer Program provided industry with free technology help on a large number of support activities and useful products.

The support activities included providing equipment such as fire engines, medical help, and a search for parts for local communities. The government provided technical sensors, such as infra-red sensors, to review waste material into the local rivers.

6.14.1 Technology Transfer Stories

1. Communication System:

The Warminster Naval Air Development Center designed and developed a Communication System for mentally handicapped person. The system basically provided 576 pictures, with a hand touch switching system, that allowed the handicapped person to identify a picture or symbol that will convey what they could not convey through language or writing. The system was designed and developed by John Mudryk, Tom Depasqua and Jerry Miller.

2. Trolley Car and Light Rail Train Simulator:

The Philadelphia Transit System wanted to build a Trolley Car and Light Rail Train Simulator system. Ray Glemser and I traveled to center city Philadelphia, in 1995 to ride a number of trolley cars throughout the city and review the problems of the trolley car travel, such as traffic problems and turning a corner with a car parked close to the trolley car travel lane. Ray and I designed a potential Trolley Car and Light Rail Train Simulator and provided the design to the Philadelphia Transit System. The Philadelphia Transit System pursued obtaining funding to build the system and as far as I know the funding was not obtained.

Chapter 7 Systems Engineering Conclusions and Recommendations

7.1 Conclusions

The main premise of this book is to present the idea that every phase of the design and development of a product should include a Systems Engineering approach. The Systems Engineering approach elements, including the hardware, software, modeling, interface, human factors, security, logistics and safety elements should be reviewed to properly coordinate each element, so that a fully successful product can be developed.

This book, in Chapters 1 through 6 presents general details of all the Systems Engineering elements and applies these elements in Chapters 8 through 14 to a design and development a specific product. The specific product, presented in Chapters 8 through 14, is the design and development of an F-18C and F-14B/A & D Flight Simulator.

This book presents the product, process and performance used to produce a successful system.

This book also presents some of the program factors that affect the success of the product, including the organization factor, management factor, people factor, Base Closure and Realignment (BRAC) factor, promotion factor, travel factor, presentation factor, ethics factor, sexual harassment factor, personal problems factor and change factor.

This book also provides many stories of Systems Engineering incidents that possibly caused design and development problems. These stories should be read and analyzed to understand the lessons learned to provide a successful product.

7.2 Recommendations

The primary recommendation is to provide the product process with a complete Systems Engineering approach to ensure that the hardware, software, modeling, interface, human factors, security, logistics and safety factor have been properly coordinated. This process should apply a single page approach of analysis of each System Engineering element. Presented below are a series of one page Product Analysis suggestions for each Systems Engineering elements.

7.2.1 Hardware Product Analysis Review

The Hardware Product Analysis List is presented below, in Table 7.2.1.1

Table 7.2.1.1 Hardware Product Analysis List	
TASK NO.	**TASK**
1	Generate a list of all product components.
2	Generate a Systems Engineering Hardware Block Diagram.
3	Generate a list of computers to provide product.
4	Generate specific computer requirements, including the following - System architecture - Processing capability (MIPS) - Number of Processors - Distribution of Software Modules between Processors - Internal Cache Capability (KB) - External Cache Capability (MB) - RAM Capability (GB) - Hard Drive Capability (GB)
5	Exercise critical software routines on the computer before the Critical Design Review (CDR)

7.2.2 Software Product Analysis Review

The Software Product Analysis List is presented below, in Table 7.2.2.1

Table 7.2.2.1 Software Product Analysis List	
TASK NO	**TASK**
1	Identify software language
2	Identify list of software subroutines
3	Identify software architecture
4	Generate a Systems Engineering Software Block Diagram
5	Provide a software lines of code estimate
6	Provide a software cost estimate
7	Provide a distribution of software subroutines to each computer

7.2.3 Modeling Product Analysis Review

The Modeling Product Analysis List is presented below, in Table 7.2.3.1

Table 7.2.3.1 Modeling Product Analysis List	
TASK NO	**TASK**
1	Generate the program model types, such as aircraft, weapons, radar, jammers, etc.
2	Generate a list of specific program models
3	Generate a Systems Engineering Model Block Diagram
4	Identify model architecture
5	Identify modeling inputs and outputs
6	Identify model software language
7	Identify model rules required, such as tactics rules, skill level rules, visual rules, etc.
8	Identify fidelity requirements for each model
9	Generate the model operational requirements
10	Generate the verification and validation performance requirements

7.2.4 Interface Product Analysis Review

The Interface Product Analysis List is presented below, in Table 7.2.4.1

Table 7.2.4.1 Interface Product Analysis List	
TASK NO	**TASK**
1	Generate a list of the program interface components.
2	Generate a Systems Engineering Interface Block Diagram.
3	Generate a list of the program specific computer interface components.
4	Identify the computer specific interface requirements - Bus type - Bus architecture - Bus bandwidth (MB/sec) - Bandwidth between computer external cache and processor (MB/sec) - Bandwidth between computer RAM and processor (MB/sec) - Bandwidth between computer hard drive and processor (MB/sec)

7.2.5 Human Factors Product Analysis Review

The Human Factors Product Analysis List is presented below, in Table 7.2.5.1

TASK NO	**TASK**
Table 7.2.5.1 Human Factors Product Analysis List	

1
 	
Analysis
	Analyze the program Hardware, Software, Modeling, Interface, Security, Logistics and Safety Systems Engineering elements to provide the following operation
	
	   - Achieve maximum performance by operator, control and maintenance Personnel
	   - Minimize skill and personnel operations and training time
	   - Achieve reliability of personnel equipment
	   - Foster system design standardization

2 Plan	Generate a program Human Factors Plan to address the Hardware, Software, Modeling, Interface, Logistics and Safety Systems Engineering elements to provide the operation identified in Task 1.
3 Procedure	Generate a program Human Factors Procedure to address the Hardware, Software, Modeling, Interface, Logistics and Safety Systems Engineering elements to provide the operation identified in Task 1.
4 Mitigation	Generate a program Mitigation Process to address the Hardware, Software, Modeling, Interface, Logistics and Safety Systems Engineering elements to those elements that do not seem to be providing proper Human Factors operation.

7.2.6 Security Product Analysis Review

The Security Product Analysis List is presented below, in Table 7.2.6.1

Table 7.2.6.1 Security Product Analysis	
TASK NO	**TASK**
1 Analysis	Analyze the program Hardware, Software, Modeling, Interface, Human Factors, Logistics and Safety Systems Engineering elements to provide the proper Security classification of each questionable item
2 Plan	Generate a program Security Plan to address the program Hardware, Software, Modeling, Interface, Human Factors, Logistics and Safety Systems Engineering elements to provide the protection of the classified items
3 Procedure	Generate a program Security Procedure to address the program Hardware, Software, Modeling, Interface, Human Factors, Logistics and Safety Systems Engineering elements to provide the protection of the classified items.
4 Mitigation	Generate a program Mitigation Process to address the program Hardware, Software, Modeling, Interface, Human Factors, Logistics and Safety Systems Engineering elements that do not seem to have security protection.

7.2.7 Logistics Product Analysis Review

The Logistics Product Analysis List is presented below, in Table 7.2.7.1

Table 7.2.7.1 Logistics Product Analysis List	
1 Analysis	Analyze the program Hardware, Software, Modeling, Interface, Human Factors, Security and Safety Systems Engineering elements to provide the required documentation, program material parts, spare parts, program maintenance, program testing, and configuration management.
2 Plan	Generate a program Logistics Plan to address the program Hardware, Software, Modeling, Interface, Human Factors, Security and Safety Systems Engineering elements to provide the required documentation, program material parts, spare parts, program maintenance, program testing, and configuration management.
3 Procedure	Generate a program Logistic Procedure to address the program Hardware, Software, Modeling, Interface, Human Factors, Logistics and Safety Systems Engineering elements to provide the required documentation, program material parts, spare parts, program maintenance, program testing, and configuration management.
4 Mitigation	Generate a program Mitigation Process to address the program Hardware, Software, Modeling, Interface, Human Factors, Logistics and Safety Systems Engineering elements that do not seem to provide the proper documentation, program material parts, spare parts, program maintenance, program testing, or configuration management.

7.2.8 Safety Product Analysis Review

The Safety Product Analysis List is presented below, in Table 7.2.8.1

Table 7.2.8.1 Safety Product Analysis List	
TASK NO	**TASK**
1 Analysis	Analyze the program Hardware, Software, Modeling, Interface, Human Factors, Security and Logistics Systems Engineering elements to identify the Safety of each questionable item.
2 Plan	Generate a program Safety Plan to address the program Hardware, Software, Modeling, Interface, Human Factors, Security and Logistics Systems Engineering elements to address the protection of the program Safety items.
3 Procedure	Generate a program Safety Procedure to address the program Hardware, Software, Modeling, Interface, Human Factors, Security and Logistics Systems Engineering elements to provide the protection of the specific Safety items identified.
4 Mitigation	Generate a program Mitigation Process to address the program Hardware, Software, Modeling, Interface, Human Factors, Security and Logistics Systems Engineering elements that do not seem to have safety protection.

Chapter 8 Design and Development of a Research and Trainer Flight Simulator

8.1 Introduction

A Flight Simulator is a ground based replica of the actual aircraft. Flight Simulators are used for many engineering functions, and the two main flight simulator systems are Research Flight Simulators and Trainer Flight Simulators. Research Flight Simulators are utilized in the evaluation of new aircraft hardware, aircraft software updates, aircraft human factors evaluation and conducting experimental design exercises. Trainer Flight Simulators are utilized for the training of many of the aircraft missions that would usually be pursued in an aircraft. The advantages of utilizing a flight simulator in pursuing an engineering exercise are the following,

1. Safety-Pilot / Aircraft

2. Cost- Aircraft Fuel

3. Repeatability

4. Availability-Weather Conditions

5. Data Collection and Analysis Techniques

6. Parameter Variability

7. Configuration Changes

8. Controlled Environment

A Research and Trainer Flight Simulator has been used many times to solve many engineering problems. This section of the book will detail the proper Systems Engineering Approach in the design and development of a typical Research and Trainer Flight Simulator. This will include descriptions of the overall system and the system components, including the hardware, software, modeling, interface, human factors, security, logistics, and safety.

The first issue that must be addressed when developing a Trainer Flight Simulator is "What are the Training Requirements?" Training requirements are usually in the form of training missions which, in the past, were normally accomplished using an aircraft. Some of the generic training missions are listed below.

1. Air-To-Air Intercept

2. Air-To-Ground Weapon Delivery

3. Air-To-Ship Weapon Delivery

4. Integrated Air Defense System (Surface-To-Air Weapon Sites)

5. Electronic Warfare (Jammers, Chaff and Flares)

6. Outer-Air-Battle

7. Air-Combat-Maneuvering (Dog Fight)

The Air Combat Maneuvering (Dog Fight) mission, for example, requires a 360 degree visual "field-of-view" and high fidelity aircraft, weapons and radar models. The detailed training requirements will be discussed in this section of the book.

The specific training requirements were continually asked for during the design and development of the F-14B & D Trainer and the F-18 Trainer Programs. The training requirements were never specifically provided. This is a major System Engineering omission for these programs.

The first issue that must be addressed when developing a Research Flight Simulator is "What is the specific research issue that will be addressed?" The research issues may include aircraft problems with the hardware, software, interface, human factors or safety.

8.2 Research and Trainer Flight Simulation History

From the website http://inventors.about.com/cs/ inventorsalphabet/a/ed_link.htm the following Link Trainer information is provided.

Edwin A. Link, is the inventor of the Link Trainer and founder of Link Aviation Company, in Binghamton, New York. Link received a patent on a device he called the "Pilot Maker", in 1929, which became the first ground based training flight simulator to teach pilots how to fly. Edwin Link's odyssey in becoming the "Father of Flight Simulation" began in 1927, when at age 23, he met Major Ocker at Wright Field in Ohio trying to help a group of aviators understand the problems with direction that are encountered while in flight. He would blindfold the pilots and twist them around in a seat a few times, then ask them which way they were turning. They quite often identified the wrong way, which is one of the things that gave Edwin the idea that you could make an entire aircraft simulator to train a pilot to do everything. Edwin thought you had to have an instrument that indicated where you were turning and whether you were flying straight or level.

The pilot trainer's stubby wooden cockpit fuselage was mounted on organ bellows that Link borrowed from his father's piano factory. An electric pump drove the organ bellows that allowed the trainer to bank, climb and dive as a pilot operated the controls in the cockpit. In 1933 Link added aviation instruments and radio aids and gages that could help a pilot know that he was flying level. The first significant military sale of the Link Aeronautical Trainer was in 1934 when the United States Army Air Corps purchased six trainers to develop well-trained and capable instrument pilots.

The first Link Trainer to be sold to the airline was delivered to American Airlines in 1937.

Link's simulators have trained millions of airmen, including 500,000 in the United States and allied countries during World War II.

Many other Flight Simulator Companies evolved throughout the years including the following.

1. Link Aviation Company
2. Rediffusion Simulation, Incorporated
3. Curtiss-Wright Corporation
4. Singer Simulation Corporation
5. General Electric Company
6. McDonnell-Douglas Corporation
7. Northrop Grumman Corporation
8. General Dynamics Corporation
9. Evans and Sutherland Company
10. Hughes Aircraft Company
11. Boeing Vertol Company
12. L-3 Communication, Incorporated
13. CAE Electronics LTD.
14. Thales Training & Simulation Company
15. Rockwell Collins Corporation

A fairly complete history of flight simulation is provided in the website, http://en.wikipedia.org/wiki/flight_simulator

Chapter 9 Flight Simulator Development Technical Approach

9.1 Introduction

9.2 Flight Simulator Technical Approach

The components needed in the design of a Research and Trainer Flight Simulator is presented below, in Figure 9.2.1.

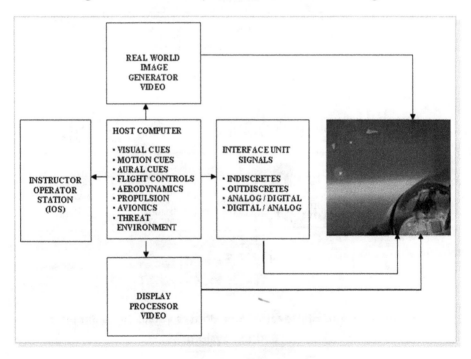

Each one of the components, identified in Figure 9.2.1, will be detailed later in various sections of the book.

9.3 Flight Simulator Exercise Technical Approach

The aircraft development phases versus flight simulator support systems, detailed in Figure 9.3.1 below, provides insight into the utilization of a flight simulator to analyze issues regarding the design and development of an aircraft system.

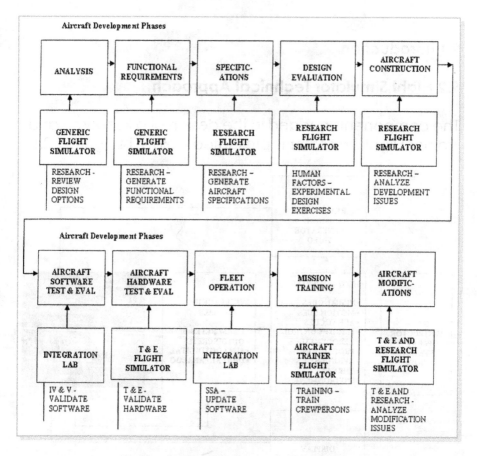

Figure 9.3.1 Aircraft Development Phases versus Flight Simulator Support Systems

The Research and Trainer Flight Simulator are utilized for the following engineering efforts.

1. Research

2. Training

3. Human Factors

4. Software Support Activity (SSA)

5. Independent Verification and Validation (IV&V)

6. Test and Evaluation (T & E)

The initial flight simulator design can be a generic architecture. As the aircraft development progresses the flight simulator design must have more of an exact replica design. The Trainer Flight Simulator needs to be an exact replica of the real world aircraft to eliminate negative training.

The Integration Lab, utilized to evaluate and update aircraft software, is built with actual aircraft avionics to ensure that the software will operate on the actual aircraft equipment. The Integration Lab is not usually a flight simulator, but simply a lab full of avionics equipment, interfaced to the actual with the aircraft computer. The Integration Lab could include a flight simulator capability.

Chapter 10 Flight Simulator Requirements

10.1 Introduction

The first issues that must be addressed when designing a Trainer and Research Flight Simulator is to identify the specific utilization of the final product. A Trainer Flight Simulator needs to identify the training requirements that will be pursued with the flight simulator. A Research Flight Simulator needs to identify the specific research effort that will be accomplished. A typical list of training requirements will be identified in the next section of this book.

The specific training requirements were continually asked for during the design and development of the F-14B & D Trainer and the F-18 Trainer Programs. The training requirements were never specifically provided. This is a major System Engineering omission for these programs.

After the training requirements are specifically determined then the hardware, software, modeling, interface, human factors, security, logistics and safety element details must be thoroughly provided. In addition, the relationship between each Systems Engineering element must be provided. The next Chapters in this book will detail information on each one of the Systems Engineering elements.

The F-18 and F-14B/A & D Trainer Flight Simulator requirements include the Operational Requirements Document (ORD) and the System Requirements Document (SRD). The F-18 and F-14B/A & D Trainer ORD and SRD will be provided later in this book.

Many sample flight simulator trainer documents are provided in the next couple of chapters. Some of the documents are part of the F-18 Trainer design and development and some of the documents are part of the F-14 B/A & D Trainer design and development. The sample document is provided depending on the one that I thought was most representative for the specific issue. Each sample document will be labled for the specific F-18 or F-14 A/B & D Trainer identified, although some of the docments such as the trainer flight simulator training missions, presented on the next page, will apply to both trainers.

10.2 Flight Simulator Training Requirements

The training requirements are usually identified in the form of aircraft missions. A list of most of the potential aircraft missions are presented below. The generic training mission name is provided in Table 10.2.1 and the technical training mission name is also identified, alone with the mission definition.

Table 10.2.1 Trainer Flight Simulator Training Missions		
Generic Mission Name	**Technical Mission Name**	**Technical Mission Definition**
1. Air-To-Air Intercept 2. Outer Air Battle (OAB) 3. Air Combat Maneuvering (A CM) [Dogfight]	1. Anti-Air Counter Air Warfare (AAW)	AAW is the destruction or neutralization of threat aircraft and combatants through the employment of air-to-air missiles and guns. The objective is to deny the opposing force the effective use of his aircraft and missile assets.
4. Air-To-Ground Weapon Delivery	2. Strike Warfare Strategic Attack / Interdiction (STW)	STW entails the destruction or neutralization of threat tactical and strategic targets ashore, near the battle area, through the employment of air-to-ground weapons, including bombs and guns. Threat targets include strategic targets, building yards, C3 centers, and operating ports and bases. The objective of Strike Warfare is to deny the opposing force the effective use of his military assets.

5. Air-To-Surface Weapon Delivery	3. Anti-Surface Warfare (ASUW)	SUW is the destruction or neutralization of threat surface combatants and merchant/logistics supply vessels through the employment of air and surface launched missiles, aircraft, surface gunfire, and electronic systems. The objective of SUW is to deny the opposing force the effective use of his surface combatants and logistic support shipping.
6. Integrated Air Defense System (IADS)	4. Suppression of Enemy Defenses (SEAD)	SEAD is the destruction or neutralization of the Surface-to-Air (SAM) and Anti-Aircraft Artillery (AAA) Sites including the missiles, radar, and command and control capability. The objective is to eliminate the opposing force's ability to damage any incoming aircraft.
7. Electronic Warfare (EW)	5. Space & Electronic Warfare (SEW)	SEW is the detection, identification, localization and targeting or neutralization of threat assets through the employment of electronic active and passive measures. Electronic and physical countermeasures are utilized to frustrate its attacks ands countermeasures to penetrate its electronic defenses. The objective of SEW is to localize and identify the opposing force electronically and deny him the effective use of his electronic and weapon assets.

8. Anti-Submarine Warfare (ASW)	6, Anti-Submarine Warfare (ASW)	ASW is the destruction or neutralization of threat submarines through the employment of air, surface and submarine assets to find and localize the threat, torpedoes and missiles to neutralize it, acoustics, electronic & physical countermeasures and decoys to confound it. The objective of ASW is to deny the opposing force the effective use of his submarine assets
9. Interactive Aircraft and Ground Site Mission		The ownship aircraft is attacking a threat aircraft when a second threat aircraft attacks the ownship. The ownship needs to address this threat. A threat aircraft flies toward a Surface-to-Air (SAM) Site. The threat then flies away from the site and a different threat aircraft flies toward the SAM Site. The SAM Site radar must now address the second threat.
	7. Mine Warfare (MW)	MW is the destruction or neutralization of threat surface platforms through the employment of air and surface launched mines and the associated denial of access to or use of a given area by hostile forces.
	8. Tactical Training Strategy "From the Sea"	"From the Sea" is the utilization of the entire Expeditionary Force (EF) to accomplish a total Navy mission, with a combination of the required sub-warfare areas.

	9. Amphibious Warfare (AMW)	AMW is the landing of marine expeditionary forces and/or special operations units on hostile territory with surface or air platforms and typically requiring the support of naval and air combatants for avoidance, detection, destruction or neutralization of threat land sea and air forces. The objective of AMW is to land forces upon and secure territory while denying the opposing force the effective use of his defensive assets.
	10. Special OPS Warfare (SPECOPS)	SPECOPS is the recovery or protection of friendly assets gathering of intelligence and/or destruction or neutralization of threat targets through the employment of air, surface and submarine assets to inset / extract specially trained and equipped personnel into/from hazardous areas.

In addition to the training mission requirements, detailed in Table 10.2.1, training curriculums may include the following testing or exercise efforts.

1. Aircraft Instrumentation Familiarization

2. Aircraft Group Exercises

3. Aircraft Jamming Exercises

4. Aircraft Carrier Landing

5. Aircraft Tactics Rules Evaluation

6. Visual Priority Rules Evaluation

7. Skill Level Rules Evaluation

Chapter 11 F-18 Aircraft and F-14 Aircraft Flight Simulator

11.1 Introduction

The two major Navy aircrafts that I have been associated with are the F-18 Aircraft and the F-14 Aircraft, so these are research and trainer flight simulators that are utilized as the examples presented in this book, for the design and development of flight simulators.

11.2 F-18 Aircraft

From the web site http://www.fas.org/man/dod-101/sys/ac/f-18.htm, the following F-18 Aircraft design details are provided.

The F-18 Aircraft Hornet, made operational in 1983, is a single and two seat twin engines, multi-mission fighter/attack aircraft that can operate from either aircraft carriers or land bases. The F-18 Aircraft fills a variety of roles, including air superiority, fighter escort, and suppression of enemy air defenses, reconnaissance, forward air control, close and deep air support and day and night strike missions. Presented below in Figure 11.2.1 is an F/A-18 Aircraft illustration.

Figure 11.2.1 F-18 Aircraft

The F-18 Aircraft length is 56 feet long, 15.3 feet high, with a wingspan of 37.5 feet and a takeoff weight of about 56,000 pounds. The aircraft can fly to an altitude of about 50,000 feet and at a maximum speed of Mach 1.8.

The F-18 Aircraft has a digital control-by-wire flight control system which provides excellent handling qualities, and provides exceptional maneuverability which allows the pilot to concentrate on operating the weapons system. A solid thrust-to-weight ratio and superior turn characteristics combined with energy sustainability, enable the

F-18 Aircraft to hold its own against any adversary. The power to maintain evasive action is what many pilots consider the finest trait. In addition, the F/A-18 Aircraft is also the Navy's first tactical aircraft to incorporate digital bus architecture for the entire system's avionics suite. The benefit of this design feature is that the F-18 Aircraft has been relatively easy to upgrade on a regular, affordable basis. The only characteristic found to be marginally adequate is the F-18 Aircraft range, although the Hornet has an intercept radius of over 400 miles without external fuel tanks. In the air-to-ground role, the F-18C Aircraft can attack targets over 550 miles away and deliver conventional bombs, precision munitions, air-to-surface missiles, cluster weapons and rockets with accuracy.

The F-18A Aircraft (single seat) and the F-18B Aircraft (dual seat) became operational in 1983. The F-18A Aircraft has an APG-65 Radar. There were 400 F-18A &B Aircraft provided.

The F-18C Aircraft (single seat) and F-18D Aircraft (dual seat) became operational in 1987. The F-18C Aircraft is provided with two General Electric F404-GE-402 Enhanced Performance Engines, and is provided with an APG-73 Radar. The F-18C & D Aircraft is provided the Advanced Medium Range Air-to-Air Missile (AMRAAM), the infrared imaging Maverick Air-to-Ground missile, a navigational forward looking infrared (FLIR) pod, a raster head-up display, night vision goggles, a digital color moving map and an independent multipurpose color display.

The F-18C & D Aircraft has a synthetic aperture ground mapping radar with a doppler beam sharpening mode to generate ground maps. This ground mapping capability permits crews to locate and attack targets in adverse weather and poor visibility and to precisely update the aircraft's location relative to targets during the approach, a capability that improves bombing accuracy.

The Navy announced on 18 May 1998 that the East Coast F/A-18 squadrons will locate to the Naval Air Station Oceana in Virginia Beach, Virginia from the Naval Air Station Cecil Field in Jacksonville, Florida. This was ordered in 1995 Base Realignment and Closure Commission. A total of 156 F-18 Aircraft moved to NAS Oceana.

The F-18E & F Aircraft Super Hornet strike fighter is an upgrade of the combat-proven F-18C & D. The F-18E & F Aircraft length is 60 feet long, 15.8 feet high, with a wingspan of 44.7 feet and a maximum takeoff weight of about 66,000 pounds. The aircraft can fly to an altitude of about 50,000 feet and at a maximum speed of Mach 1.8. The F-18E & F Aircraft has been extensively redesigned with a lengthened fuselage, 25% larger wings, enlarged tail surfaces, enlarged leading edge root extensions for better high angle-of-attack performance. The F-18E & F-18F will provide the battle group commander with a platform that has a longer range than the F-18C & D. The aircraft has eleven weapon stations and can carry the complete complement of "smart" weapons, including the newest joint weapons, such as the JDAM and JSOW. The F-18E & F Aircraft has been updated to provide an Advanced Targeting Forward Looking Infrared (ATFLIR) system.

The Navy is planning to purchase 548 F-18E & F Aircraft and possibly as many as 1000 aircrafts in the future.

The F-18G Aircraft Growler will be provided as a replacement for the EA-6B Aircraft Prowler for escort and close-in jamming.

From the following website http://www.fas.org/man/dod-101/sys/missile/aim-9,aim-7, the F-18 Aircraft armament is detailed.

The F-18 Aircraft armament include a cannon for close-in encounters, sidewinder infrared homing missiles for short range encounters, Sparrow semi-active radar homing missiles for intermediate range encounters, and AMRAAM missiles for long range encounters. All of these weapons are directed and controlled by the F-18 Aircraft radar fire control system.

The 20mm 6-barrel cannon, with 520 rounds of ammunition, are internally mounted in the nose.

The F-18 Aircraft usually carries four AIM-9 Sidewinder missiles. The AIM-9 Sidewinder is 9.4 feet long, has a wingspan of 25 inches and a diameter of 5 inches. The missile has four tail fins on the rear, with a "rolleron" at the tip of each fin. These "rollerons" are spun at high speed by the slipstream in order to provide roll stability. The missile is steered by four canard fins mounted in the forward part of the missile just behind the infrared seeker head. The Sidewinder missile has a launch weight of about 180 pounds, and a maximum effective range of about 10 miles.

The F-18 Aircraft usually carries two AIM-7 Sparrow missiles. The AIM-7 Sparrow is 12 feet long and has a launch weight of about 500 pounds. The missile has two sets of delta shaped fins. There is a set of fixed fins at the rear of the missile and a set of movable fins at the middle of the missile for steering. The AIM-7M missile, introduced in 1982, featured an inverse processed digital monopulse seeker which was more difficult to detect and jam and provided better look-down, shoot-down capability.

From the website http://www.fas.org/man/dod-101/ sys/missile/aim-120.htm, the following AMRAAM missile information is obtained.

The F-18 Aircraft usually carries two AIM-120 AMRAAM missiles. The AIM-120 AMRAAM advanced medium range air-to-air missile is a new generation air-to-air missile. It has an all-weather, beyond visual range supersonic, aerial intercept, guided missile capability. The AMRAAM Missile employs active radar target tracking, proportional navigation guidance, and active radio frequency, target detection. It employs active, semi-active, and inertial navigational methods of guidance to provide an autonomous launch and leave capability against single and multiple targets in all environments.

The AMRAAM missile weighs 340 pounds and uses an advanced solid-fuel racket motor to achieve a speed of Mach 4 and a range in excess of 30 miles. In long range engagements AMRAAM heads for the target using inertial guidance and receives updated target information via data link from the launch aircraft. It transitions to a self-guiding terminal mode when the target is within range of its own monopulse radar set. With its sophisticated avionics, high closing speed, and excellent end game maneuverability, chances of escape from the AMRAAM missile are minimal.

11.3 F-18 Aircraft Flight Simulator

Enclosed, in Figure 11.3.1, is an illustration of the F-18C Aircraft Trainer Flight Simulator.

Figure 11.3.1 F-18 Aircraft Trainer

11.4 F-14 Aircraft

From the web site http://www.fas.org/man/dod-101/sys/ac/f-14.htm the following F-14 Aircraft design details are provided.

The F-14 Aircraft Tomcat is a supersonic, twin-engine, variable sweep-wing; two place fighters designed to attack and destroy enemy aircraft at day or night and in all weather conditions. The F-14 Aircraft can track up to 24 targets simultaneously. The F-14 Aircraft is illustrated below in Figure 11.4.1.

Figure 11.4.1 F-14 Aircraft

The F-14 Aircraft, provided in the early 1970's, is a two seat, twin-engine fighter with twin tails and variable geometry wings. The aircraft length is 61 feet, 9 inches long, 16 feet high, with a wingspan of 64 feet and a weight of about

72,900 pounds. The aircraft can fly to an altitude of 53,000 feet and at a maximum speed of Mach 1.88. The US Navy ordered a total of 699 F-14A Aircraft. In addition, 79 F-14A Aircraft were delivered to Iran before the 1979 revolution.

The F-14 Aircraft general arrangement consists of a long nacelle containing the large nose radar and two crew positions extending well forward and above the widely spaced engines. The engines are parallel to a central structure that flattens towards the tail; butterfly-shaped airbrakes are locked between the fins on the upper and lower surfaces. Altogether, the fuselage forms more than half of the total aerodynamic lifting surface.

The wings are shoulder mounted and are programmed for automatic sweep during flight, with manual override provided. The twin, swept fin and rudder vertical surfaces are mounted on the engine housings and canted outward. The wing pivot carry-through structure crosses the central structure; the carry through is 22 feet long and constructed from 33 electron welded parts machined from titanium; the pivots are located outboard of the engines. Normal sweep range is 20 to 68 degrees with 75 degrees over sweep position provided for shipboard hangar stowage; sweep speed is 7.5 degrees per second. For roll control below 57 degrees, the F-14 Aircraft uses spoilers located along the upper wing near the trailing edge in conjunction with its all-moving, swept tailplanes, which are operated differentially; above 57 degrees sweep, the tailplanes operate alone. For unswept, low speed combat maneuvering, the outer two sections of the trailing edge flaps can be deployed at 10 degrees and the nearly full-span leading-edge slats are drooped to 8.5 degrees. At speeds above Mach 1.0, glove vanes in the leading edge of the fixed portion of the wing extend to move the aerodynamic center forward and reduce loads on the tailplane.

The F-14A Aircraft, introduced in the mid-1970s, is the basic platform of the F-14 Aircraft series. It is equipped with two Pratt & Whitney TF30-P-414A engines. The F-14A Aircraft utilizes an AWG-9 Radar. The aircraft cockpit front seat area includes a head-up display, vertical situation display, a horizontal situation display, television camera set, radar warning receiver, chaff / flare dispenser and a deception jamming pod. The F-14 Aircraft updates, throughout the years, included the ALR-67 Countermeasure Warning and Control System, Forward Looking Infrared (LANTIRN FLIR) System, Programmable Tactical Information Display (PTID), Digital Flight Control System (DFCS) and the Tactical Air Reconnaissance Pod System (TARPS).

The F-14B Aircraft is equipped with two General Electric F110-GE-400 engines to replace the F-14A Aircraft Pratt & Whitney TF30-P-414A engines. The new engine emphasizes reliability, maintainability and operability, which improves capability and maneuverability without throttle restrictions or engine trimming. There were 157 F-14 B Aircrafts produced.

The F-14D Aircraft is equipped with two F110-GE-400 engines, an APG-71 Radar to replace the AWG-9 Radar. The APG-71 Radar replaces the analog processing hardware with a flexible digital processing radar. The F-14D Aircraft upgrades the aircraft avionics with a new signal processor, data processor, digital display, Joint Tactical Information Distribution System (JTIDS), and an Infrared Search and Track System (IRST), which provides long range detection in the long wave infrared spectrum of both subsonic and supersonic targets. There were 53 F-14D Aircraft developed.

From the following website http://www.fas.org/man/dod-101/sys/missile/aim-9,aim-7, the F-14 Aircraft armament is detailed.

The F-14 Aircraft armament include a cannon for close-in encounters, sidewinder infrared homing missiles for short range encounters, Sparrow semi-active radar homing missiles for intermediate range encounters, and Phoenix missiles for long range encounters. All of these weapons are directed and controlled by the F-14 Aircraft radar fire control system.

The 20mm cannon are carried on the port side of the forward fuselage. A total of 675 rounds are carried in a drum and when the guns are fired the empty cases are returned to the drum rather than being ejected overboard.

The F-14 Aircraft usually carries four AIM-9 Sidewinder missiles. The AIM-9 Sidewinder is 9.4 feet long, has a wingspan of 25 inches and a diameter of 5 inches. The missile has four tail fins on the rear, with a "rolleron" at the tip of each fin. These "rollerons" are spun at high speed by the slipstream in order to provide roll stability. The missile is steered by four canard fins mounted in the forward part of the missile just behind the infrared seeker head. The Sidewinder missile has a launch weight of about 180 pounds, and a maximum effective range of about 10 miles.

The F-14 Aircraft usually carries two AIM-7 Sparrow missiles. The AIM-7 Sparrow is 12 feet long and has a launch weight of about 500 pounds. The missile has two sets of delta shaped fins. There is a set of fixed fins at the rear of the missile and a set of movable fins at the middle of the missile for steering. The AIM-7M missile, introduced in 1982, featured an inverse processed digital monopulse seeker which was more difficult to detect and jam and provided better look-down, shoot-down capability.

The F-14 Aircraft usually carries two AIM-54 Phoenix missiles. The AIM-54 Phoenix missile is 13.2 feet long, the body is 13 inches wide, and the wing span is 3 feet. The launch weight is about 985 pounds, and the maximum missile range is about 90 miles. The Phoenix missile is propelled by a single –stage solid fuel rocket motor, which gives a maximum velocity up to Mach 5, at high altitudes. The missile has four fixed delta-shaped wings and is steered by tail-mounted control surfaces. After launch, the Phoenix can use three different types of guidance, including autopilot, semi-active radar homing, and fully-active radar homing. For long range shots, the missile generally flies a pre-programmed route immediately after launch under autopilot control. At mid-course, the nose-mounted radar seeker takes over, operating in semi-active mode, honing in on radar waves reflected off the target from the F-14 Aircraft radar system. Once it gets within about 14 miles of the target, the Phoenix own radar takes over for the final run in to the target, and the missile operates in the fully-active radar homing mode.

In addition, the F-14 Aircraft, in 1994, was provided an air-to-ground capability, including the MK-82, MK-83 and MK-84 Bombs, MK-20 Cluster Bombs and GBU-10, GBU-12, GBU-16 and GBU-24 Laser Guided Bomb.

The Navy announced on 17 February 2006 that the last F-14 Aircraft mission was completed on 8 February 2006. A pair of Tomcats landed aboard the USS Theodore Roosevelt aircraft carrier after one F-14 Aircraft dropped a bomb in IRAQ. The F-14 Aircraft is being replaced by the F-18E/F Super Hornet. The F-14 Aircraft entered the Navy service in September 1974 with a maximum speed of 1,584 mph costing $47 million. Since 1974 the F-14 Aircraft performed many important missions and in 1986 its Navy pilots were at the heart of the "Top Gun" movie in which Tom Cruise played Maverick, an impetuous pilot training at the Navy's elite flight school in Miramar, California.

11.5 F-14 Aircraft Flight Simulator

Enclosed, in Figure 11.5.1, is an illustration of the F-14 Aircraft Trainer Flight Simulator.

Figure 11.5.1 F-14 Aircraft Trainer

Chapter 12 Flight Simulator Products

12.1 Introduction

Systems Engineering transforms an operational need into a final product. The Research and Trainer Flight Simulator product requires a System Engineering approach to produce a successful product. The flight simulator need is usually identified by a group of people that are very familiar with the operation of research and trainer flight simulators. In the Department of Defense these people are usually the military that best understand the operation of the real world systems that are going to be simulated. The need and the general flight simulator that will solve the need are then documented. In the government this document is currently the Operational Requirements Document (ORD), which is usually generated by the technical management, such as NAVAIR, in the Navy. The Operational Requirements Document (ORD) report background is presented in Section 12.2 of this book. The Operational Requirements Document (ORD) is then expanded with technical details in the System Requirements Document (SRD). The System Requirements Document (SRD) report background is presented in Section12.3 of this report.

The development of any research or trainer flight simulator involves a tradeoff of the product cost, product schedule and the product quality. The estimate of the product cost, schedule and quality of the product is very difficult to determine and is usually based on the past experience of the development of past similar research and trainer flight simulator products. The better the initial product analysis, the generation of the flight simulator requirements, and particularly the training or research requirements, the more accurate the product cost, schedule and quality will be. As the product development progresses, the cost, schedule and quality of the product usually needs to be updated to provide the true values of each System Engineering element.

The major factor of all programs, including the development of a research or trainer flight simulator is funding. When you are out of money the program ends.

Many managers are more concerned with the cost and schedule of the product, while engineers are more concerned with the quality of the product. Many managers say that only two of the three factors can be obtained, but in my opinion, the cost, schedule and quality of the product should be analyzed in a System Engineering procedure, and tradeoff pursued to obtain the best product possible.

In the Department of Defense, there is a number of funding type that is used to produce a variety of flight simulator products. Presented below, in Table 12.1.1, is the Department of Defense funding types and utilization.

Table 12.1.1 Department of Defense Funding Types	
Funding Type	**Funding Utilization**
APN	Aircraft / Flight Simulator Procurement
APN5	Aircraft / Flight Simulator Modification
APN6	Aircraft / Flight Simulator Spares and Repair Parts
APN7	Aircraft / Flight Simulator Facility Support Equipment
OMN	Aircraft / Flight Simulator Operation and Maintenance
OMNR	Aircraft / Flight Simulator Reserve Operation and Maintenance
OPN	Aircraft / Flight Simulator Other Procurements
PANMC	Aircraft / Flight Simulator Ammunition
RDT & E	Aircraft / Flight Simulator Research, Development, Test & Evaluation
WPN	Aircraft / Flight Simulator Weapons Procurement
WPNMOD	Aircraft / Flight Simulator Weapon Modifications

The Department of Defense currently uses the Systems Application & Products in Data Processing (SAP), which is the leading Enterprise Resource Planning (ERP) software provider in the world. SAP is a German software company, started by two former IBM Company workers.

Presented below, in Figure 12.1.1, is the Department of Defense funding planning procedure.

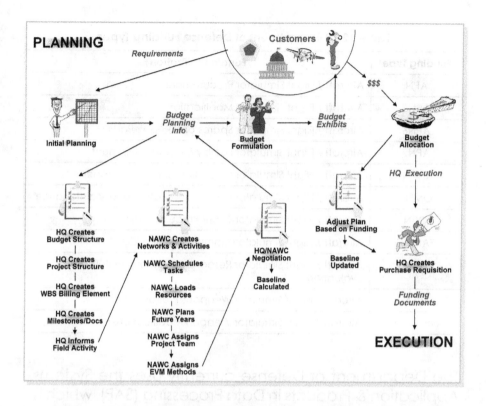

Figure 12.1.1 Department of Defense Funding Planning Procedure

Presented below, in Figure 12.1.2, is the Department of Defense funding execution procedure.

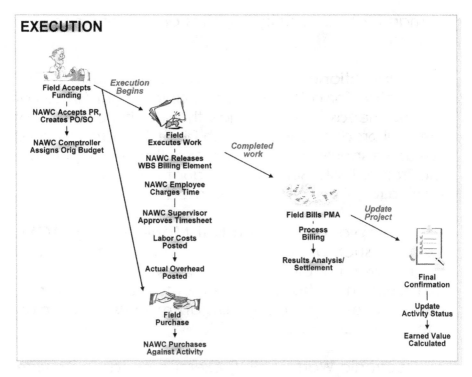

EXECUTION

Field Accepts Funding
|
NAWC Accepts PR, Creates PO/SO
|
NAWC Comptroller Assigns Orig Budget

Execution Begins

Field Executes Work
|
NAWC Releases WBS Billing Element
|
NAWC Employee Charges Time
|
NAWC Supervisor Approves Timesheet
|
Labor Costs Posted
|
Actual Overhead Posted

Completed work

Field Purchase
|
NAWC Purchases Against Activity

Field Bills PMA
|
Process Billing
|
Results Analysis/ Settlement

Update Project

Final Confirmation
|
Update Activity Status
|
Earned Value Calculated

Figure 12.1.2 Department of Defense Funding Execution Procedure

12.2 Operational Requirements Document (ORD)

The Operational Requirements Document (ORD) provides the product need and the general system that provide the need. The Operational Requirements Document (ORD) Report outline is provided in Section 3.1.1 of this book.

The most important thing to be considered when generating the Operational Requirements Document (ORD) is that all of the Systems Engineering elements, including the hardware, software, modeling, interface, human factors, security, logistics and safety, need be discussed.

The Operational Requirements Document (ORD) generated for a Research or Trainer Flight Simulator product needs to fully provide the specific utilization of the final product. The Research Flight Simulator product needs the specific research exercises that will be pursued. The Trainer Flight Simulator product needs the specific training requirements to be detailed.

For example, an Air Combat Maneuvering (ACM) exercise, such as a "dog fight", needs to have a 360 degree visual display field-of-view, which probably will be an expensive hardware element. The ACM exercise also requires high fidelity aircraft models and a high fidelity tactics model.

An Electronic Warfare (EW) exercise needs to provide high fidelity jammer models, which could be costly.

The Operational Requirements Document (ORD) will identify the approximate cost and schedule for producing the Research or Trainer Flight Simulator product.

It is also important that the managers decide whether the Research or Trainer Flight Simulator product will be produced in-house or on a contract to an outside contractor. The Research or Trainer Flight Simulator products are usually contracted out. The contract package generated uses the Operational Requirements Document (ORD) and the System Requirements Document (SRD) as the contract Statement of Work.

12.3 System Requirements Document (SRD)

The System Requirements Document (SRD) report expands on the Operational Requirements Document (ORD) to provide the specific technical requirement details that allows the design and development team to provide the product to accomplish the product need. The System Requirement Document (SRD) Report outline is provided in Section 3.1.2 of this book.

The most important thing to be considered when generating the System Requirement Document (SRD) is that all of the Systems Engineering elements, including the hardware, software, modeling, interface, human factors, security, logistics and safety, need be discussed.

The System Requirements Document (SRD) generated for a Research or Trainer Flight Simulator product needs to fully provide the specific utilization of the final product. The Research Flight Simulator product needs the specific research exercises that will be pursued. The Trainer Flight Simulator product needs the specific training requirements to be detailed.

A typical System Requirements Document (SRD), such as a portion of the F-14B/A Trainer Program, Naval Air Warfare Center, Training Division, Orlando, Florida, is presented below. The F-14B/A Trainer SRD is numbered from 1.0 to 6.2 and the portion of the report presented below provides the exact numbering of the F-14B/A Trainer SRD.

System Requirements Document (SRD) Sample Section

3.1.4 The F-14B/A Weapon Systems Trainer Device 2F169 shall consist of the following systems,

3.1.4.1 Trainee Station System Function: The trainee station shall provide a complete set of controls and displays for the trainee pilot and RIO which replicates those of the F-14B cockpit of the design basis aircraft. These controls and displays, as well as the physical components which represent the cockpit environment, such as seats and canopy, shall be functionally integrated with the trainer subsystems which support the trainer simulation. The trainer shall respond to activation of controls. The trainer shall replicate the display data as a function of the conditions and status of the simulation. The flight characteristics, control responses, instrument readings, and warning, caution, and advisory lights stimulus shall be simulated.

3.1.4.2 Instructor / Operator Station (IOS) System Function: The IOS shall provide the controls, displays and related facilities to control, monitor and evaluate the trainer and trainee operations. The IOS shall provide the components and features for presentation, control, and monitoring of training exercises, including: system control and status, problem status, problem control, communications, mode control and clocks. The IOS shall control malfunctions of avionics, aircraft systems, and cockpit instruments and circuit breakers.

3.1.4.3 Not used

3.1.4.4 Visual System Function: A visual system shall provide the aircrew with real time, out-the-window displays of day, night and dark visual scenes correlated with both the G-cuing and sensor functions. The visual system shall include scenes for flight takeoff and landing for both land and ship conditions.

3.1.4.5 G-Cuing System Function: The G-Cuing system shall provide the aircrew rapid and highly responsive onset cues and sustained cues. These cues shall be correlated with both visual and sensor functions.

3.1.4.6 Aural Cuing System Function: The aural cuing shall provide the aircrew with realistic sounds associated with aircraft operation.

3.1.4.7 Sensor Simulation System Function: The sensor simulation shall provide the aircrew with sensor cues associated with navigation, targets and weapon delivery.

3.1.4.8 Digital Computation System (DCS) Function: The DCS shall provide the capability to program, generate and control the training device system.

3.1.4.9 Digital Conversion Equipment (DCE) System Function: The DCE system shall convert digital information stored in the computation system to formats usable by the trainer subsystems and shall convert data from the trainee station to formats unstable by the DCS and other trainer subsystems.

3.1.4.10 Power System Function: The Power System shall provide complete electrical power to the training device.

The F-14B/A Weapon Systems Trainer Device 2F169 shall consist of the following system function relationships and performance.

3.1.5 System Functional Relationships: The performance, flying qualities, aircraft systems, electronic equipment, and instrument response and control reactions of the design basis aircraft shall be simulated in accordance with this specification. Functional relationships between various systems of the aircraft shall be included in the simulation except as specified. Controls and indicators which appear in the aircraft cockpit shall be provided in the trainer and perform as specified herein.

3.1.5.1 Simulation Performance: The trainer shall provide real time simulation of normal, degraded and emergency aircraft operation with respect to transient and steady state flight, engine performance, flying qualities, communication and navigation systems operations, environmental effects, tactical operations, ground operations, and flight path. Such simulation shall be reflected by the display, instrument, signal, and aural indications, control reactions, and flight path display traces, responding to trainee and instructor control inputs. Representative system errors, characteristics and anomalies shall be simulated. The trainer shall simulate the operation and performance of the aircraft, equipment, subsystems, control, and instruments, in the modes of operation. The trainer shall simulate the design basis aircraft performance and all of its systems throughout the flight envelope. In the event of conflict between this specification and the design basis aircraft data, the design basis aircraft data shall take precedence.

The following are some of the F-14B/A Trainer SRD details of the trainer Aircraft Carrier.

3.1.5.1.7 Aircraft Carrier: The aircraft carrier CVN-68 class shall be simulated. The simulation shall provide carrier motion, aircraft launch and recovery equipment, electronic equipment, landing and taxi director, and a visual display of the carrier. A three dimensional representation of the carrier shall be used. Data for the carrier model shall be obtained from the CV NATOPS Manual, and the aircraft launch and recovery bulletins. Instructor controls shall be provided for functions and launch and recovery performed by shipboard personnel. Visual simulation of the carrier and the Fresnel Lens Optical Landing System (FLOLS) shall be provided in accordance with the specific specification.

3.1.5.1.7.1 Flight Deck: The carrier simulation shall depict the flight deck consisting of the landing area, lighting, arresting cables, barricade, deckedge and catapults. The deck edge shall be simulated as a series of straight lines within five feet of the deck edge required by the virtual scene. The island shall be modeled as a rectangular solid for purposes of crash detection. The abrupt change in radar altitude and the loss of ground effect at the deck edge shall be simulated.

3.1.5.1.7.2 Carrier Motion: Carrier motion shall be simulated for pitch, roll, yaw, heave, direction and speed throughout the ranges specified. The carrier's motion shall be made a function of the carrier's performance characteristics, the sea state, wind direction and speed. Controls shall be provided for the instructor to reset the position of the carrier within the ocean gaming area. Position of the carrier shall be dynamically simulated.

3.1.5.1.7.3 Carrier Aerodynamic Effects: Aerodynamic effects of the carrier shall be simulated for turbulence and carrier burble.

3.1.5.1.7.4 Carrier Take-off: The complete sequence of aircraft launch shall be simulated for engine start-up, taxi on deck, connecting the launch bar and holdback, tensioning the catapult and launch. The lights associated with the integrated catapult control system shall be simulated. Catapult acceleration of the aircraft shall be based on aircraft gross weight and excess and speed as input from the instructor console.

3.1.5.1.7.5 Carrier Landing: Carrier landing shall be simulated for wave-off, touchdown point, normal arrestment, and run out deceleration. Hook position shall be dynamically positioned and used to determine if an arrestment occurs, however, simulation of hook bounce is not required. Four arresting wires shall be simulated.

Presented below are some details of the gaming area and the tactical environment.

3.1.6.3.1 Tactical Gaming Area: Tactical gaming areas shall be selectable by the instructor / operator.

The tactical gaming area shall encompass air space from ground and sea level to an altitude of 150,000 feet with a ground range of 2000 nm by 2000 nm. Calculations for the tactical gaming area shall be in accordance with the atmospheric and geographic simulation through magnetic variation. Control of the tactical environment shall be accomplished at the IOS console.

3.1.6.3.2 Threats: Threats shall consists of the following,

- Air Platform (Aircraft)
- Surface Platforms (Ships)
- Ground Platforms
- Ground Sites
- Air-to-Air Missiles
- Air-to-Surface Missiles
- Air-to- Ground Bombs
- Surface-to-Air Missiles
- Surface-to-Surface Missiles
- Decoys

A site is the simulated location of a system represented by a jammer, radar, missile system, missile launch complex, and fire control system which remains fixed during the simulated scenario.

A platform is the simulation of ground, surface and air vehicles associated with a system represented by a jammer, radar, missile system and fire control system which have the capability to move during the simulated scenario.

It shall be possible to detect these threats by the appropriate ownship sensors as defined in the sensor paragraphs specified herein. Sensor detection of threats shall replicate that achieved with the operational equipment aboard the design basis aircraft. Probability of detection shall be based on threat range aspect, electromagnetic emission, day or night meteorological conditions and position from ownship.

Targets are defined as a threat which can be challenged, engaged and destroyed by ownship. Targets shall be a subset of threats and contain the above except AAM and SAM missiles. Emitters shall consist of jammers, radar systems and electromagnetic guidance systems. The simulation shall correlate active emitters with threats.

Provisions shall be provided to place the following number of targets and features within a scenario at a location during both the plan modes and during the execution of a scenario,

Aircraft	96
Missiles	32
Ground (Moving)	12
Ground (Fixed)	48
SAM / AAA	48
Surface (Ships)	48
Weather Patterns	12

The instantaneous and simultaneous target and feature requirements are specified in the threat and individual subsystem paragraphs. The visual system shall dynamically compute which targets to display to the aircrew as a result of range, position and its threat priority to ownship.

12.4 Flight Simulator Elements (Hardware, Software, Modeling, Interface, Human Factors, Security, Logistics and Safety)

12.4.1 Hardware

The design and development of a Research and Trainer Flight Simulator requires the identification of the hardware components that can accomplish the specific program requirements. The identification of the Flight Simulation hardware details depends on the specific utilization of the system, discussed in the last section of the book. Presented below are some of the flight simulator hardware choices that need to be selected when determining the specific hardware that will satisfy the requirements.

Cockpit Configuration:

Generic
Exact Replica
Generic & Exact Replica
Combination
Integration Lab

Indicator
Aircraft Type:

Fixed Winged
Helicopter
Vertical Short Field Takeoff &Landing (V/STOL)

Seating Configuration:

Single Seat
Side by Side
Tandem
Three Crewpersons
Four Persons
Display

Cockpit Flight Controls:

Control Stick
Side Arm Controller
Collective
Throttle

Cockpit Instrumentation:

Airspeed Indicator
Barometric Altimeter

Radar Altimeter
Attitude Indicator
Horizontal Situation

Rate-of Climb Indicator
Accelerometer
Angle-of-Attack Indicator
Wing Sweep Indicator

Cockpit Display:

Multipurpose
Vertical Situation Display
Horizontal Situation Display
Digital Display
Tactical Information

Head-Up Display

Computer:

Digital
Analog
General Purpose

Rudders	Graphics Image Generator
	Graphics Work Station
Sensors:	
	Crewstation / Computer
	Interface
Radar	
Television	D / A Converter
	(Instruments)
LOFAR Buoy	A / D Converter (Flight
	Controls)
DIFAR Buoy	Indiscrete (Switches)
CASS Buoy	Outdiscrete (Lights)
Infra-Red Search & Track System (IRST)	Synchro Converter
Magnetic Anomaly Detection System (MAD)	Video
Electronic Support Measures System (ESM)	

A typical aircraft flight simulator hardware block diagram is provided in Figure 12.4.1.1. The block diagram is taken from the F-14 D Aircraft Trainer Program. Some typical flight simulator hardware components will be described in this section of the book, including cockpit crewstations, visual display systems, motion base systems, flight simulator computers and instructor operator stations.

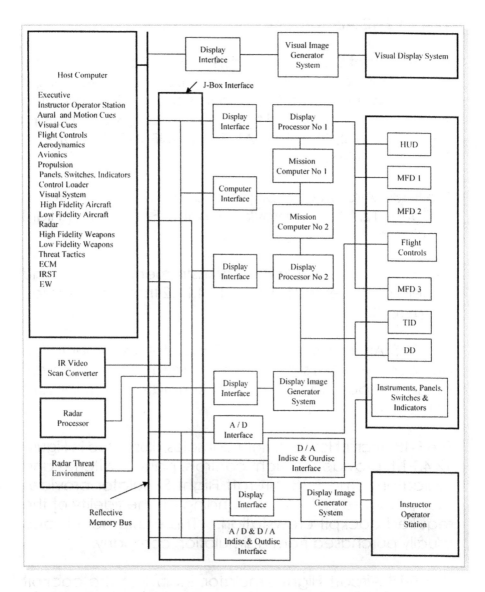

Figure 12.4.1.1 F-14D Flight Simulator Trainer Typical Hardware Block Diagram

12.4.1.1 F-18 Aircraft Cockpit Crewstation

The F-18 Flight Simulator consists of a cockpit crewstation, which includes instrumentation, displays, sensors, flight controls, panels, switches and indicators. The following illustrations are provided from the F/A-18 A/B/C/D Natops Flight Manual, Number A1-F18AC-NFM-000, dated 1 December 1985.

The F-18C Aircraft cockpit panel and consoles, illustrated in Figure 12.4.1.1.1, is provided in the F-18 Aircraft Flight Simulator.

The F-18 Aircraft cockpit list of instruments, display and panels are presented in Figure 12.4.1.1.2 that was illustrated in Figure 12.4.1.1.1

The F-18 Aircraft Trainer Flight Simulator instrumentation and displays utilize simulated equipment. The equipment is usually purchased from a simulation equipment company.

The F-18 Aircraft left and right consoles, illustrated in Figure 12.4.1.1.1, provide the flight controls, panel, switches and indicators. The F-18 Aircraft Flight Simulator provides many of these systems, depending on the fidelity of the required cockpit crewstation. This equipment is also usually purchased from a simulation company.

The F-18 Aircaft Flight Simulator, including the cockpit crewstation was designed and developed by L3 Communications Company, in Dallas, Texas. L3 Communications Company basically includes the Singer Simulation Company, formerly in Binghamton, New York and the Hughes Aircraft Company, formerly in Los Angeles, California, which were bought by L3 Communications.

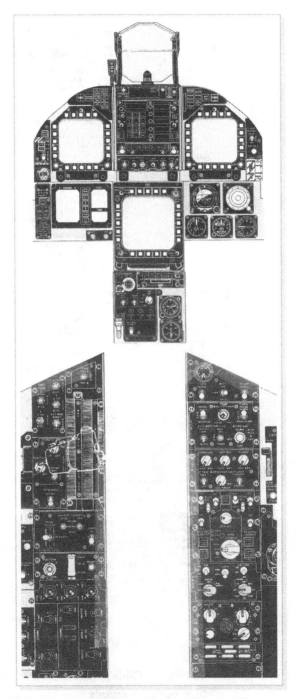

Figure 12.4.1.1.1 F-18 Aircraft Cockpit Crewstation and Consoles

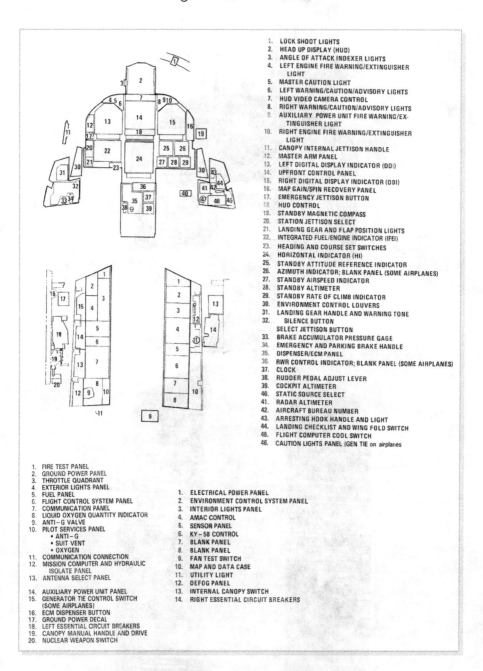

1. LOCK SHOOT LIGHTS
2. HEAD UP DISPLAY (HUD)
3. ANGLE OF ATTACK INDEXER LIGHTS
4. LEFT ENGINE FIRE WARNING/EXTINGUISHER LIGHT
5. MASTER CAUTION LIGHT
6. LEFT WARNING/CAUTION/ADVISORY LIGHTS
7. HUD VIDEO CAMERA CONTROL
8. RIGHT WARNING/CAUTION/ADVISORY LIGHTS
9. AUXILIARY POWER UNIT FIRE WARNING/EXTINGUISHER LIGHT
10. RIGHT ENGINE FIRE WARNING/EXTINGUISHER LIGHT
11. CANOPY INTERNAL JETTISON HANDLE
12. MASTER ARM PANEL
13. LEFT DIGITAL DISPLAY INDICATOR (DDI)
14. UPFRONT CONTROL PANEL
15. RIGHT DIGITAL DISPLAY INDICATOR (DDI)
16. MAP GAIN/SPIN RECOVERY PANEL
17. EMERGENCY JETTISON BUTTON
18. HUD CONTROL
19. STANDBY MAGNETIC COMPASS
20. STATION JETTISON SELECT
21. LANDING GEAR AND FLAP POSITION LIGHTS
22. INTEGRATED FUEL/ENGINE INDICATOR (IFEI)
23. HEADING AND COURSE SET SWITCHES
24. HORIZONTAL INDICATOR (HI)
25. STANDBY ATTITUDE REFERENCE INDICATOR
26. AZIMUTH INDICATOR; BLANK PANEL (SOME AIRPLANES)
27. STANDBY AIRSPEED INDICATOR
28. STANDBY ALTIMETER
29. STANDBY RATE OF CLIMB INDICATOR
30. ENVIRONMENT CONTROL LOUVERS
31. LANDING GEAR HANDLE AND WARNING TONE SILENCE BUTTON
32. SELECT JETTISON BUTTON
33. BRAKE ACCUMULATOR PRESSURE GAGE
34. EMERGENCY AND PARKING BRAKE HANDLE
35. DISPENSER/ECM PANEL
36. RWR CONTROL INDICATOR; BLANK PANEL (SOME AIRPLANES)
37. CLOCK
38. RUDDER PEDAL ADJUST LEVER
39. COCKPIT ALTIMETER
40. STATIC SOURCE SELECT
41. RADAR ALTIMETER
42. AIRCRAFT BUREAU NUMBER
43. ARRESTING HOOK HANDLE AND LIGHT
44. LANDING CHECKLIST AND WING FOLD SWITCH
45. FLIGHT COMPUTER COOL SWITCH
46. CAUTION LIGHTS PANEL (GEN TIE on airplanes

1. FIRE TEST PANEL
2. GROUND POWER PANEL
3. THROTTLE QUADRANT
4. EXTERIOR LIGHTS PANEL
5. FUEL PANEL
6. FLIGHT CONTROL SYSTEM PANEL
7. COMMUNICATION PANEL
8. LIQUID OXYGEN QUANTITY INDICATOR
9. ANTI-G VALVE
10. PILOT SERVICES PANEL
    • ANTI-G
    • SUIT VENT
    • OXYGEN
11. COMMUNICATION CONNECTION
12. MISSION COMPUTER AND HYDRAULIC ISOLATE PANEL
13. ANTENNA SELECT PANEL
14. AUXILIARY POWER UNIT PANEL
15. GENERATOR TIE CONTROL SWITCH (SOME AIRPLANES)
16. ECM DISPENSER BUTTON
17. GROUND POWER DECAL
18. LEFT ESSENTIAL CIRCUIT BREAKERS
19. CANOPY MANUAL HANDLE AND DRIVE
20. NUCLEAR WEAPON SWITCH

1. ELECTRICAL POWER PANEL
2. ENVIRONMENT CONTROL SYSTEM PANEL
3. INTERIOR LIGHTS PANEL
4. AMAC CONTROL
5. SENSOR PANEL
6. KY-58 CONTROL
7. BLANK PANEL
8. BLANK PANEL
9. FAN TEST SWITCH
10. MAP AND DATA CASE
11. UTILITY LIGHT
12. DEFOG PANEL
13. INTERNAL CANOPY SWITCH
14. RIGHT ESSENTIAL CIRCUIT BREAKERS

Figure 12.4.1.1.2 F-18 Aircraft Cockpit List of Instruments, Displays and Panels

12.4.1.2 F-14 Aircraft Cockpit Crewstation

The F-14 Flight Simulator consists of a cockpit crewstation, which includes instrumentation, displays, sensors, flight controls, panels, switches and indicators. The F-14 Aircraft has a front and back seat configuration. . The following illustrations are provided from the F-14A Natops Flight Manual, Number NAVAIR 01-F14AAA-1, dated 1 January 1980.

The F-14A Aircraft front seat cockpit panel and consoles, illustrated in Figure 12.4.1.2.1, is provided in the F-14A Aircraft Flight Simulator. The F-14A Aircraft Flight Simulator provides many of these systems, depending on the fidelity of the required cockpit crewstation.

The F-14A Aircraft front cockpit list of instruments, displays and panels are, presented below in Figure 12.4.1.2.2, that was illustrated in Figure 12.4.1.2.1.

The F-14A Aircraft back seat cockpit panel, is illustrated in Figure 12.4.1.2.3, and is provided in the F-14A Aircraft Flight Simulator.

The F-14A Aircraft back seat cockpit list of instruments, displays and panels are presented in Figure 12.4.1.2.4, that was illustrated in Figure 12.4.1.2.3.

The F-14A Aircraft Flight Simulator provides many of these systems, depending on the fidelity of the required cockpit crewstation.

The F-14 Aircraft Trainer Flight Simulator instrumentation and displays utilize simulated equipment. My favorite simulated instrument company is the Malwin Electronics Corporation in Paterson, New Jersey and my favorite simulated display company is the Electronic Image Systems, Inc in Xenia, Ohio.

The F-14D Aircraft Trainer Flight Simulator was designed and developed by McDonnell Douglas Corporation, in St Louis Missouri and AAI Corporation in Cockeysville, Maryland.

The F-14B/A Aircraft Trainer Flight Simulator was designed and developed by Grumman Aerospace Corporation, in Great River, New York.

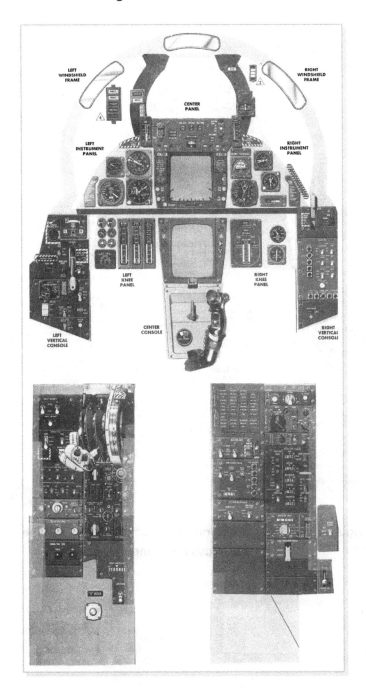

Figure 12.4.1.2.1 F-14 Aircraft Front Panel and Consoles

Figure 12.4.1.2.2 F-14 Aircraft Front Cockpit List of Instruments, Displays and Panels

LEFT SIDE CONSOLE
1. G VALVE PUSHBUTTON
2. OXYGEN-VENT AIRFLOW CONTROL PANEL
3. COMM/NAV COMMAND CONTROL PANEL
4. INTEGRATED CONTROL PANEL
4a UHF (AN/ARC 159)
4b UHF COMM SELECT PANEL
5. TONE VOLUME CONTROL PANEL
6. ICS CONTROL PANEL
7. AFCS CONTROL PANEL
8. THROTTLE QUADRANT
9. INLET RAMPS/THROTTLE CONTROL PANEL
10. TARGET DESIGNATE SWITCH
10a. HYDRAULIC HAND PUMP

LEFT VERTICAL CONSOLE
11. FUEL MANAGEMENT PANEL
12. CONTROL SURFACE POSITION INDICATOR
12a LAUNCH BAR ABORT PANEL
13. LANDING GEAR CONTROL PANEL
14. WHEELS FLAPS POSITION INDICATOR
14a. EMER STORES JETTISON BUTTON

LEFT KNEE PANEL
15. ENGINE PRESSURE RATIO INDICATOR
16. EXHAUST NOZZLE POSITION INDICATOR
17. OIL PRESSURE INDICATOR
18. HYDRAULIC PRESSURE INDICATOR
19. ELECTRICAL TACHOMETER INDICATOR (RPM)
20. THERMOCOUPLE TEMPERATURE INDICATOR (TIT)
21. RATE OF FLOW INDICATOR (FF)

LEFT INSTRUMENT PANEL
22. RADAR ALTIMETER
23. SERVOPNEUMATIC ALTIMETER
24. AIRSPEED MACH INDICATOR

25. VERTICAL VELOCITY INDICATOR
26. LEFT ENGINE FUEL SHUT OFF HANDLE
27. ANGLE-OF-ATTACK INDICATOR

LEFT FRONT WINDSHIELD FRAME
28. APPROACH INDEXER
29. WHEELS WARNING LIGHT
29a. BRAKES WARNING LIGHT
30. ACLS/AP CAUTION LIGHT
30a. NWS ENGA CAUTION LIGHT

CENTER PANEL
31. HFADS UP DISPLAY
32. AIR COMBAT MANEUVER PANEL
33. VERTICAL DISPLAY INDICATOR (VDI)
34. HORIZONTAL SITUATION DISPLAY INDICATOR (HSI)
35. PEDAL ADJUST HANDLE
36. BRAKE PRESSURE INDICATOR
37. CONTROL STICK

RIGHT FRONT WINDSHIELD FRAME
38. ECM WARNING LIGHTS
39. STANDBY COMPASS

RIGHT INSTRUMENT PANEL
40. WING SWEEP INDICATOR
41. RIGHT ENGINE FUEL SHUT OFF HANDLE
42. ACCELEROMETER
43. STANDBY ATTITUDE INDICATOR
44. CANOPY JETTISON HANDLE
45. CLOCK
46. BEARING DISTANCE HEADING INDICATOR (BDHI)
47. UHF REMOTE INDICATOR

RIGHT KNEE PANEL
48. FUEL QUANTITY INDICATOR
49. LIQUID OXYGEN QUANTITY INDICATOR
50. CABIN PRESSURE ALTIMETER

RIGHT VERTICAL CONSOLE
51. ARRESTING HOOK PANEL
52. DISPLAYS CONTROL PANEL
53. ELEVATION LEAD PANEL

RIGHT SIDE CONSOLE
54. COMPASS CONTROL PANEL
55. CAUTION—ADVISORY INDICATOR
56. TACAN CONTROL PANEL
57. MASTER GENERATOR CONTROL PANEL
58. ARA-63 CONTROL PANEL
59. AIR CONDITIONING CONTROL PANEL
60. MASTER LIGHT CONTROL PANEL
61. EXTERNAL ENVIRONMENTAL CONTROL PANEL.
62. MASTER TEST PANEL
63. HYDRAULIC TRANSFER PUMP SWITCH
64. DEFOG CONTROL LEVER
65. WINDSHIELD DEFOG SWITCH

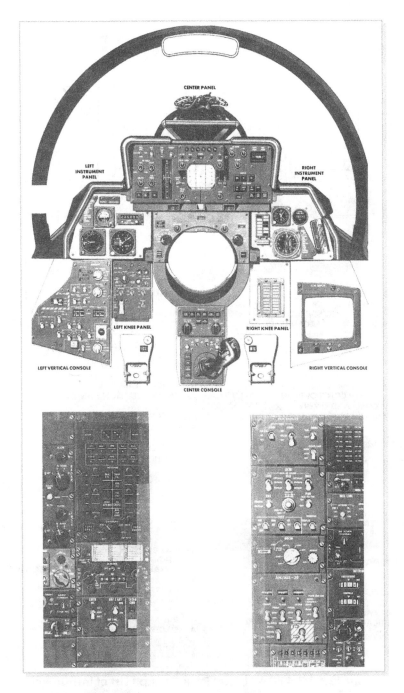

Figure 12.4.1.2.3 F-14 Aircraft Rear Cockpit Panel and Consoles

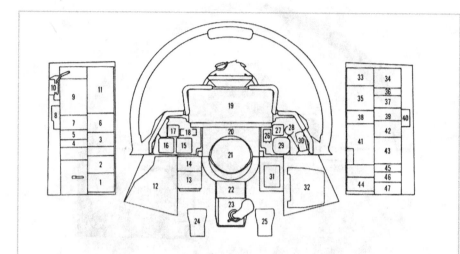

LEFT SIDE CONSOLE
1. G VALVE PUSHBUTTON
2. OXYGEN-VENT AIRFLOW CONTROL PANEL
3. XMTR SEL/ANT SEL/TACAN CMD
4. KY-28 CONTROL PANEL
5. ICS CONTROL PANEL
6. ARC 51 UHF CONTROL PANEL
   ARC 159A (V)5 CONTROL PANEL ⚠1
7. TACAN CONTROL PANEL
8. LIQUID COOLING CONTROL PANEL
9. RADAR IR/TV CONTROL PANEL
10. EJECT COMMAND LEVER
11. COMPUTER ADDRESS PANEL

LEFT VERTICAL CONSOLE
12. ARMAMENT PANEL

LEFT KNEE PANEL
13. SYSTEM POWER PANEL
14. SYSTEM TEST PANEL

LEFT INSTRUMENT PANEL
15. SERVOPNEUMATIC ALTIMETER
16. AIRSPEED MACH INDICATOR
17. STANDBY ATTITUDE INDICATOR
18. UHF REMOTE INDICATOR

CENTER PANEL
19. DETAIL DATA DISPLAY PANEL (DDD)

CENTER CONSOLE
20. NAVIGATION CONTROL AND DATA READOUT
21. TACTICAL INFORMATION DISPLAY (TID)
22. TACTICAL INFORMATION CONTROL PANEL
23. HAND CONTROL UNIT

LEFT AND RIGHT FOOT WELLS
24. ICS FOOT BUTTON
25. MIC FOOT BUTTON

RIGHT INSTRUMENT PANEL
26. THREAT ADVISORY LIGHTS
27. CLOCK
28. FUEL QUANTITY TOTALIZER
29. BEARING DISTANCE HEADING
    INDICATOR (BDHI)
30. CANOPY JETTISON HANDLE

RIGHT KNEE PANEL
31. CAUTION-ADVISORY PANEL

RIGHT VERTICAL CONSOLE
32. MULTIPLE DISPLAY INDICATOR

RIGHT SIDE CONSOLE
33. ECM DISPLAY CONTROL PANEL
34. DIGITAL DATA INDICATOR (DDI)
35. ECM CONTROL PANEL
36. DATA LINK CONTROL PANEL
37. DATA LINK REPLY AND ANT CONTROL PANEL
38. DECM CONTROL PANEL
39. AA1 CONTROL PANEL
40. DEFOG CONTROL LEVER
41. AN/ALE-39 PROGRAMMER AND CONTROL
42. INTERIOR LIGHT CONTROL PANEL
43. IFF TRANSPONDER CONTROL PANEL
44. MID COMPRESSION BYPASS TEST PANEL
45. IFF ANTENNA CONTROL AND TEST PANEL
46. RADAR BEACON CONTROL PANEL
47. ELECTRICAL POWER SYSTEM TEST PANEL

Figure 12.4.1.2.4 F-14 Aircraft Rear Cockpit List of Instruments and Displays

12.4.1.3 Flight Simulator Visual Display System Specifications

The flight simulator "out-the-window" visual display system provides the visual cues needed to conduct many of the program exercises. The selection of the visual display system that will best accomplish the job depends on the specific research or trainer requirements. For example, an Air Combat Maneuvering (Dog Fight) requires a 360 degree field-of-view.

Presented below are typical flight simulator visual display system requirement parameters that will provide for a number of mission exercises.

Field-of-View:	180 degrees horizontal, 60 degrees Vertical, with a 30 degree downward view
Resolution:	5 Arc Minutes
Brightness:	5 Ft Lamberts
Contrast:	20 to 1
Exit Pupil Diameter:	4 Ft (Minimum)
Eye Relief:	4 Ft (Minimum)
Picture Color:	Red/Blue/Green Chromatic

1. Flight Simulator Visual Display System Field of View

The field-of-view is defined as the maximum cone of rays passed by the aperture of the system and measured at a given vertex, which is the pilot's eye. The field-of-view requirements vary for each aircraft being simulated and according to the mission of the aircraft.

2. Flight Simulator Visual Display System Resolution

Resolution is the ability of an optical system to transfer detail faithfully. Consider an optical system which images two equally bright point sources of light. When the separation is such that it is just possible to determine that there are two points and not one, the points are said to be resolved.

The measure of resolution for an optical system is described as line pairs per milli-meter of resolution. Television resolution is generally defined as the number of "TV Lines" alternate stripes of black and white visible on the horizontal dimensions of the picture equal to the vertical dimension. This television resolution definition does not consider the distance of the viewer from the screen although; in actually it is a considerable factor. The optical system resolution unit also does not consider distance.

The human eye is capable of resolving 1 arc minutes of resolution. Since this resolution unit is an angular unit, the distance from the viewer to the picture is not required.

Most real world visual simulations utilize television systems for their visual displays. Nine "TV Lines" are required for image recognition, according to The Design Analysis Report, NAVTRAEQUIPC-1H-197, Naval Equipment Center, Orlando, Florida, December 1972.

3. Flight Simulator Visual Display System Brightness

When luminous flux strikes a surface, the surface is said to be illuminated. The luminance is defined as the luminous flux incident per unit area. The unit of luminance is the foot-candle or ft lamberts, which is 1 lumen incident per square foot. The ft lamberts are the brightness which results from one foot candle of illumination falling on a perfect diffusing surface. In modern illuminating engineering practice luminance is expressed in lumens per square meter. Some typical values of luminance are;

Sunlight: 100,000 lumens / meter^2 (9300 ft lamberts)

Close Work: 100 lumens / meter^2 (9.3 ft lamberts)

The lumen is defined as the luminous flux emitted into a solid angle of one steradian by a point source of light. A steradian is the solid angle subtended by ¼ pi of the surface area of a sphere. A sphere subtends 4 pi steradians.

A simple method of comprehending the value of a steradian is that it subtends one square foot of surface area on a one foot diameter sphere. Since one lumen is incident on the one square foot area under an illumination of one foot-candles, the total flux radiated into a hemisphere of 2 pi steradians from a perfectly diffuse surface is just one lumen.

4. Flight Simulator Visual Display System Depth-of-Field

When a lens is focused to give a sharp image of a particular object, other objects closer or farther away do not appear equally sharp. The decline of sharpness is gradual and there is a zone extending in front of and behind the focused distance where the blur is too small to be noticeable and can be accepted as sharp. This zone is known as the depth of field of the lens.

In an optical pickup in which a television link is employed to convert the optical image to a convenient display, the depth of field is considerably limited as compared with real life situations. The ability of an observer's eyes to adapt to a large range of focal distances in the real life situation cannot be utilized in the simulation display because of the fixed focal plane of the television camera tube.

5. Flight Simulator Visual Display System Contrast

Contrast is the range of light and dark values in a picture or the ratio between the minimum and maximum vales of brightness. Contrast is an important parameter because a picture with good contrast and only fair resolution sometimes appears sharper than a picture with good resolution and only fair contrast.

6. Flight Simulator Visual Display System Exit Pupil Diameter

The exit pupil diameter is defined as that opening (real or virtual) which represents the common base of all the cones of light emerging from the system. Another way to define the exit pupil is the image of the aperture stop of the system as formed by the optics. The larger the exit pupil, the greater the lateral head movement that is possible while still being able to see the full field-of-view. For a side by side cockpit configuration an especially large exit pupil diameter is required, possibly 48 inches, if both pilot and co-pilot are to see the same visual display.

7. Flight Simulator Visual Display System Eye Relief

Eye relief is defined as the distance from the pilot's eye to the closest optical element. Normally, as a minimum, the eye relief should be the distance from the pilot's eye to the cockpit instrument panel, about 30 inches. The maximum dimension depends on the cockpit geometry / motion system constraints, probably about 10 feet. The eye relief dimension should be as large as possible to accommodate a wide variety of aircraft.

8. Flight Simulator Visual Display System Picture Color

Chromatic color has been proven necessary in flight simulation through numerous experiments. Realistic chromatic color provides identification information for known objects, such as water, earth, trees, airfield lights, etc. In addition, color in visual systems enhances the apparent contrast of objects. This feature is important in a search and acquisition mission.

12.4.1.4 Flight Simulator Visual Display Hardware

Presented below are a number of typical Flight Simulator Visual Display System hardware systems.

1. Front Projection Screen Visual Display System

The Front Projection Screen Visual Display System is illustrated below in Figure 12.4.1.4.1

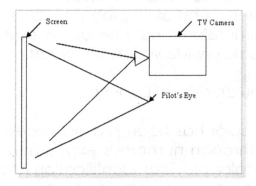

Figure 12.4.1.4.1 Front Projection Screen Visual Display

The Projection Screen Visual Display System is the earliest method used in a flight simulator. This system utilizes a large screen front or rear projection screen onto which the scene is projected. The screen is generally from 6 to 10 feet from the pilot's eye and the visual effect is the same as seen in a movie theater. Since the image presented is a real image and only 6 to 10 feet away, the eyes are not focused at infinity so a true real world simulation of eye focusing is not obtained.

The advantage of the Projection Screen Visual Display System method is that it is simple in theory, simple to implement, low in cost and a real image produced without optical distortions.

2. Dome Front Projection Screen Visual Display System

The Dome Front Projection Screen Visual Display System is illustrated below in Figures 12.4.1.4.2 and 12.4.1.4.3

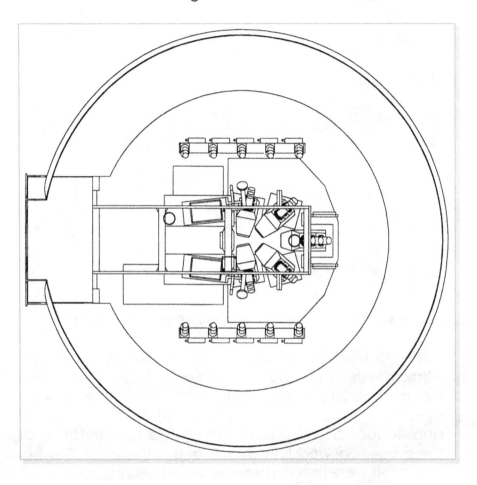

Figure 12.4.1.4.2 F-14B Trainer Dome Visual Display System (Top View)

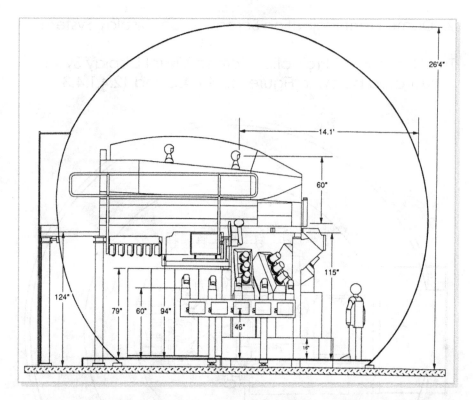

Figure 12.4.1.4.3 F-14B Trainer Dome Visual Display System (Side View)

The F-14B Trainer utilized a Front Projection Screen Visual Display System, in a 30 foot dome and a 40 foot dome. If a real image is placed 15 feet away from the viewer's eye, the image is assumed to be at infinity, so the 30 and 40 foot dome system eliminated the limitation of the a true real world simulation of eye focusing. Several projectors were used to provide a low fidelity background, fairly high fidelity forward insets, and high fidelity threat images.

Presented below are the single Dome Projection Screen typical specifications.

Field-of-View:

Background-	320 degrees horizontal, 140 degrees vertical
3 Inserts-	220 degrees horizontal, 75 degrees vertical
Targets-	20 degrees horizontal, 20 degree vertical

Resolution:

Background-	32 arc minutes
3 Insets-	8.5 arc minutes
Targets-	1 arc minute

Brightness:

Background-	0.5 ft lamberts
3 Insets-	2 ft lamberts
Targets-	3 Ft lamberts

Picture Color: Red/Blue/Green Chromatic

3. Rear Projection Screen Visual Display System

The Rear Projection Screen Visual Display System is illustrated below in Figure 12.4.1.4.4

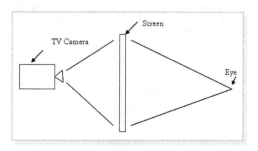

Figure 12.4.1.4.4 Rear Projection Screen Visual Display

The Rear Projection Screen Visual Display System, presented in Figure 12.4.1.4.4, is similar to the Front Projection Screen, used in a flight simulator. This system utilizes a large screen rear projection screen onto which the scene is projected, utilizing the projector behind the screen. The screen is generally from 6 to 10 feet from the pilot's eye and the visual effect is the same as seen in a movie theater. Since the image presented is a real image and only 6 to 10 feet away, the eyes are not focused at infinity, so a true real world simulation of eye focusing is not obtained.

The advantage of the Rear Projection Screen Visual Display System method is that it is simple in theory, simple to implement, low in cost and a real image produced without optical distortions.

4. Spherical Mirror / Beam Splitter / CRT Visual Display System

The Spherical Mirror / Beam Splitter / CRT Visual Display System is illustrated below in Figure 12.4.1.4.5

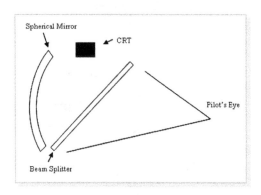

Figure 12.4.1.4.5 Spherical Mirror / Beam Splitter / CRT Visual Display System

The Spherical Mirror / Beam Splitter / CRT Visual Display System are an excellent system with in-line optics that provides a sharp, high fidelity, bright image. The field-of-view is limited, so a number of systems have to be utilized to provide an adequate field of view for most aircraft missions.

Presented below are the single Spherical Mirror / Beam Splitter / CRT Visual Display System typical specifications.

Field-of-View: 48 degrees horizontal, 32 degrees Vertical

Resolution: 6 Arc Minutes

Brightness: 4 Ft Lamberts

Contrast: 30 to 1

Exit Pupil Diameter: 6 inches

Eye Relief: 24 inches

Picture Color: Red/Blue/Green Chromatic

5. Off-Axis Visual Display System

The Off-axis Visual Display System is illustrated below, in Figure 12.4.1.4.6

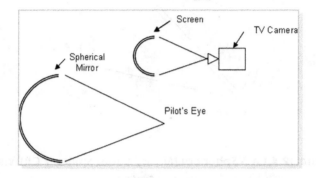

Figure 12.4.1.4.6 Off-Axis Visual Display System

The Off-Axis Visual Display System provides a larger field-of-view than the Spherical Mirror / Beam Splitter / CRT Visual Display System, but since it does not provide in-line optics the system has some optical distortion. The Off-Axis Visual Display System also incorporates more than one system to provide a larger field-of-view.

Presented below are the single Off-Axis Visual Display System typical Specifications.

Field-of-View: 60 degrees horizontal, 60 degrees Vertical

Resolution: 5 Arc Minutes

Brightness: 5 Ft Lamberts

Contrast: 20 to 1

Exit Pupil Diameter: 4 feet

Eye Relief: 4 feet Minimum

Picture Color: Red/Blue/Green Chromatic

6. Helmet Mounted Visual Display System

The Helmet Mounted Visual Display System is illustrated in Figure 12.4.1.4.7

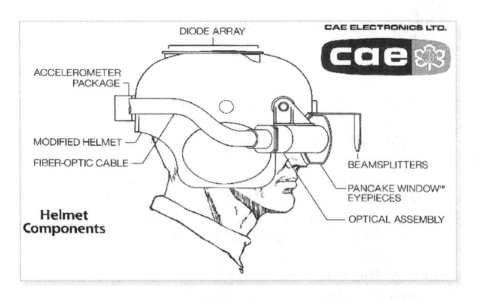

Figure 12.4.1.4.7 Helmet Mounted Display System

The CAE Helmet Mounted Visual Display System consists of a lightweight helmet with imagery transferred to a small display provided in front of the pilot's eyes. An optical helmet tracking system provides head position information, while an accelerometer package, mounted on the rear of the helmet, and provides data for accurate lead predication to compensate for the computer image generator (CIG) transport delays.

Presented below are the Helmet Mounted Display System Specifications.

Field-of-View: 135 degrees horizontal, 64 degrees Vertical

Resolution: 4.6 Arc Minutes

Brightness: 30 Ft Lamberts

Contrast: 30 to 1

Exit Pupil Diameter: 4 feet

Picture Color: Red/Blue/Green Chromatic

12.4.1.5 Flight Simulator Image Generator System

The flight simulator "out-the-window" visual display image generator provides the visual cues needed to conduct many of the program exercises. Technical improvements to visual display system presentations has gone from using model board presentations to computer generated image presentations from the early 1970's to the computer image generator presentations today. Many of the computer image generator presentations are very detailed.

12.4.1.6 Flight Simulator Image Generator Hardware

In the early 1970's the visual display image generator utilized the gantry system optical probe and television camera to view a model board. The system is illustrated in Figures 12.4.1.6.1 and 12.4.1.6.2. This system was fairly detailed, but very limited in the gaming area.

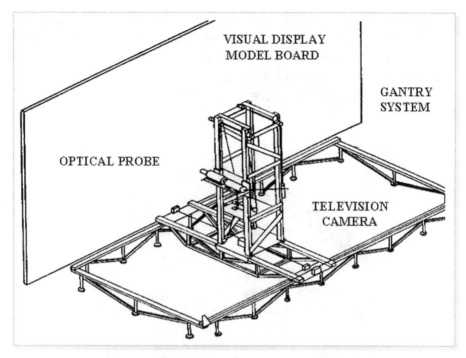

Figure 12.4.1.6.1 Visual Display System Model Board System

Figure 12.4.1.6.2 Visual Display System Model Board

The aircraft target generator, used in the 1970's before computer generator images were developed, included a hardware aircraft with a television camera system as illustrated in Figure 12.4.1.6.3. This system was also very limited in the gaming area available.

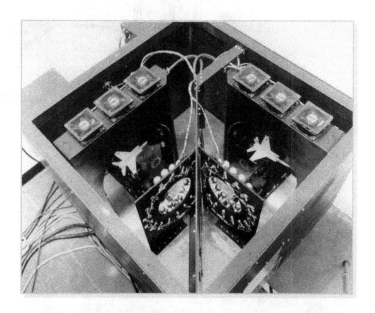

Figure 12.4.1.6.3 Aircraft Target Generator System

An early 1970's computer generated display system is provided in Figure 12.4.1.6.4 . The presentation is somewhat cartoon like.

Figure 12.4.1.6.4 1970's Computer Generated Image Display

Computer generated image presentations provide a very detailed real world image that allows an almost infinite gaming area. Some of the image generator system presentations are provided below in Figures 12.4.1.6.5 and 12.4.1.6.6 Many of the illustrations presented below are from Evans & Sutherland Computer Corporation

Figure 12.4.1.6.5 Computer Generated Image Display

Figure 12.4.1.6.6 Computer Generated Image Display

A computer generated image of an aircraft is presented below in Figure 12.4.1.6.7

Figure 12.4.1.6.7 Computer Generated Image of Aircraft Presentation

A computer generated image of a nighttime scene is presented below in Figure 12.4.1.6.8

Figure 12.4.1.6.8 Computer Generated Image of Night Time Presentation

Presented below, in Figure 12.4.1.6.9, is an illustration of a cockpit crew station with a visual display system

Figure 12.4.1.6.9 Cockpit Crew station with a Visual Display System

The methodology utilized to create a computer generated image is illustrated below, in Figure 12.4.1.6.10 from a report entitled "The Computer Image Generation", by Logion, Inc., for Wright-Patterson Air Force Base.

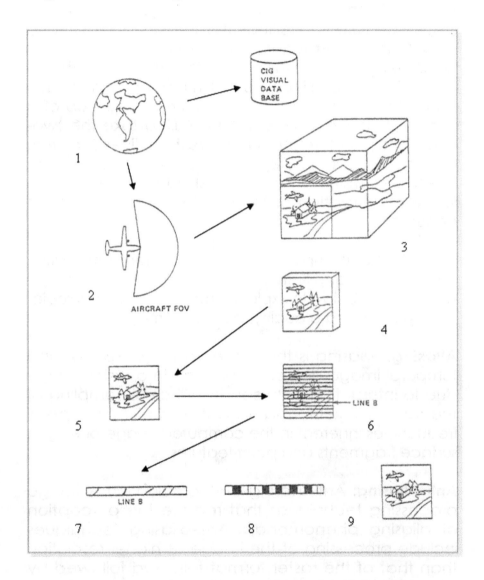

Figure 12.4.1.6.10 Computer Image Generation Methodology

Step 1 converts the real world terrain into the CGI visual data base, using the MIL-D-89020 Digital Terrain Elevation Data (DTED) and Digital Feature Analysis Data (DFAD), which include the terrain elevations and the terrain features, such as buildings, bridges, etc. Steps 2 and 3 are to reduce the amount of visual data to only the information visible through the aircraft windows. Steps 4 and 5 are to reduce the three-dimensional scene is converted into a two-dimensional scene for presentation. The picture is made up of a matrix of dots and pixels. Step 6 is to take the two-dimensional picture and subdivide it into lines of picture elements of pixels. Step 7 and 8 takes each picture line to determine its color and shade. Finally, in Step 9, the construction of the flight simulator picture is generated.

A number of computer image generator terms, taken from an Evans and Sutherland Image Generator Report, are provided below to explain some of the methodologies used in providing a quality real world image.

Aliasing: Aliasing is the appearance of spatial and temporal image defects or artifacts in a raster image, due to interaction between the discrete sampling of the raster / pixel format and the spatial / temporal frequencies inherent in the computed image of edges, surface fragments and point features.

Anti-Aliasing: Anti-aliasing is a combination of image processing techniques that reduce the perception of aliasing phenomena. Anti-aliasing techniques include processing of the image at higher resolution than that of the raster format followed followed by image filtering to reduce both spatial and temporal aliasing.

Data Base: A data base is the numerical data representing the information derived from maps, photographs and drawing of airports, terrain, water, moving objects, encoded in a digital format compatible with the image generator.

Flicker: Flicker is perceptible temporal variation in luminance.

Light Point: A light point is a simulated light emitting radiator at a specific location in the scene.

Lines per Frame: The lines per frame mean the total number of lines drawn in one frame time.

Modeling: Modeling refers to the process by which a digital data base is created from maps, charts, engineering drawings, aerial photographs, or other data.

Object: An object is a collection of polygons and / or light strings with fixed visual priority. Objects represent the structural primitives for building large data bases, and also assist in the culling steps involved with scene processing.

Occultation: Occultation is the visual obstruction of scene lights or surfaces by other surfaces. Occultation may appear as the partial or total hiding of given light points or polygons in the model by other polygons, in response to the position of the observer.

Pixel: A pixel is the smallest element of resolution along a display scanline of a raster.

Polygon: A polygon is a planar convex surface bounded by three or four defining vertices or points and the implied bounding edge segments connecting the ordered vertices. Color, reflectance and texture represent other defining attributes of a polygon.

Raster: A raster is the structure of parallel scanlines covering the usable display screen area and forming the picture.

Refresh Rate: Refresh rate is the frequency at which the image fields are re-drawn to preclude image flicker at a given scene brightness and to provide fresh eye stimulus for perception of motion cues.

Texture: Texture is the enhancement of a polygon by applying intensity modulated 2-dimensional pattern or combination of patterns to the surface.

Transport Time Delay: The transport time delay is the time required to propagate data from the initiation of a visual system cycle with time corrected control input from the host computer, to the start of the display of that field.

Update Rate: Update rate is the frequencies at which new perspective images of the scene are computed in order convey the proper impression of scene motion.

12.4.1.7 Flight Simulator Motion Base Hardware

From a report entitled "Pilot Cuing" by Dr. Edward A. Stark the following flight simulator information is provided.

The pilot is an information processor and system controller. He develops skill as a pilot by learning to recognize information which reflects significant aspects of the flight environment and by practicing and control of the relationships reflected by this information. Simulators used for training portray task and learning relevant information to support the development of perceptual, psychomotor and judgmental skills required in flight. Psychomotor or perceptual motor processes are the means by which specific muscular actions are employed in response to specific perceived information in maintaining or achieving some prescribed performance criterion. Simulators are limited in their ability to portray as much information as is available in actual flight, but effective learning is possible when critical elements of the real flight environment can be provided.

Pilot control cues are mediated by several different kinds of sensory systems. The visual system is the most important, but internal sensors in the inner ear, skin, muscles, and joints all contribute to the patterns of information needed by the pilot in flight control and learning of flight control skills. These cues are provided using a motion base system. The inner ear contains three sets of acceleration sensors which respond to angular and linear motions. Three semicircular canals in each inner ear are oriented to respond to angular rotations, each responding to motions in one of the three body planes. Pressure receptors throughout the body serve the function of assisting in the control of body and limb position with minimum visual involvement. In flight they serve the function of providing information about changes in the pressure of the aircraft seatbelt and shoulder harness on the body as changes occur in aircraft acceleration and attitude.

The two main flight simulator motion base systems are the synergistic motion base system and the centrifuge motion base system, which will be described in the next two pages of this book.

The main difficultly in the utilization of a motion base system is the coordination of the motion base system with the visual display system. Not providing a close coordination of visual and motion cues causes the pilot potential sickness.

The Navy and the Air Force have not been using a motion base system with a fixed wing aircraft flight simulator, but a motion base system is usually provided with a helicopter or vstol aircraft flight simulator. The centrifuge motion base system included the F-14 Aircraft Flight Simulator to study the high g capability during aircraft spin exercises.

The fixed wing aircraft flight simulator provides a G-Seat, G-Suit and a Vibration System to simulate some of the motion cues provided in aircraft flight. The G-Seat, G-Suit and Vibration System technical details are provided below from the "Motion and Force Cuing" report by Frank Cardullo.

The G-Seat provides sustained acceleration cues. The G-Seat induces the illusion of acceleration by providing the somatic stimuli, such as slim pressure changes and body position cues which seem to be related to the whole body acceleration. For example, as the pilot of a high performance aircraft exercises a "pull-up" maneuver, they sense an apparent increase in their body weight for the duration of the maneuver. The design of a G-Seat is such that the seat pan, back rest, and lap belt act in concert to simulate the simulator pilot's perceptual systems to create the illusion of sustained acceleration. Cuing algorithms have been developed to establish the shape and elevation of the G-Seat in order to adjust the pressure gradient across the back and buttocks of the pilot consistent with the particular maneuver being executed.

The G-Suit provides a strong cue indication of the acceleration profile of the aircraft being simulated. The G-Suit system is a low pressure pneumatic system utilizing air from the auxiliary air supply, which regulates the pressure into the pilot's G-Suit as a function of increasing G forces. It provides pilot protection against grayout and blackout, forestalls the diminution of mental alertness which may result from repeated high acceleration forces, and provides the pilot with a very good indication of the acceleration to which the aircraft is being subjected. An accurate simulation greatly enhances the realism of the training environment and provides a significant sustained acceleration cue used by the pilot in control of their aircraft.

The Vibration System is designed to supplement the vibration cues provided by platform motion systems. The range of frequencies provided by these systems is from 3 to 20 hertz. Seat shaker systems are often used to provide the vibrations. One seat shaker system produces stall buffets, Mach buffets and other vibratory cues. Another seat shaker system, used in helicopter flight simulators, provides vibrations for once per revolution of the helicopter rotors.

One of the main flight simulator motion base systems is the Synergistic Motion Base System, illustrated below in Figure 12.4.1.7.1.

Figure 12.4.1.7.1 Synergistic Motion Base System

The Synergistic Motion Base System, as provided by the Singer Link Company, includes a moving platform assembly, driven by six identical 48 inch stroke six degree-of-freedom hydraulic actuators. The motion system is capable of providing pitch, roll, yaw, lateral, longitudinal and vertical movement, either simultaneously or in any combination desired. The major motion base system components include the Hydraulic System, the Electronic System and the Moving Platform.

The Motion Base performance requirements are presented below.

1. Pitch

Rotation:	+ 31 deg. to - 28 deg.
Velocity:	+ / - 15 deg. / sec.
Acceleration:	+ / -50 deg. / sec. / sec.

2. Roll

Rotation:	+ / - 23 deg.
Velocity:	+ / - 15 deg. / sec.
Acceleration:	+ / - 50 deg. / sec. / sec.

3. Yaw

Rotation:	+ / - 32 deg.
Velocity:	+ / - 15 deg. / sec.
Acceleration:	+ / - 50 deg. /sec. / sec.

4. Vertical

Translation:	+ / - 26 inch
Velocity:	+ / - 24 inch / sec.
Acceleration:	+ / - .8g

5. Lateral:

Translation:	+ / - 42 inch
Velocity:	+ / - 24 inch / sec
Acceleration:	+ / - .6g

6. Longitudinal

Translation:	+ / - 48 inch.
Velocity:	+ / - 24 inch / sec.
Acceleration:	+ / - .6g

Another flight simulator motion base system is the centrifuge motion base system, illustrated below in Figure 12.4.1.7.2.

Figure 12.4.1.7.2 Centrifuge Motion Base System

The Centrifuge Motion Base System provides a 10 ft diameter sphere mounted on a 50 ft long arm, which is rotated in a horizontal plane about the axis of a vertically mounted 4000 hp direct current motor. The centrifuge arm has a maximum angular acceleration rate of 1.78 radians / sec and a maximum velocity of 173 miles / hours. The 10 ft diameter gondola is attached to the end of the 40 ft long arm by means of a two gimbal system which permits complete rotational freedom in the pitch and roll directions. The gondola structure can support a weight of 40,000 pounds.

The following specifications are the maximum angular motion rates about the gondola's geometric center.

Roll Acceleration	6.7 radians / sec
Pitch Acceleration	9.7 radians / sec
Yaw Acceleration	10.0 radians / sec
Roll Velocity	30 RPM
Pitch Velocity	30 RPM
Yaw Velocity	30 RPM

12.4.1.8 Flight Simulator Computers

The design and development of digital computers has provided such new technical capabilities each year that it is impossible to identify a Trainer Flight Simulator computer without analyzing the current program requirements and measuring the requirements against the current computer capabilities. In years past, Trainer Flight Simulators utilized computers with UNIX operating systems, but currently the Trainer Flight Simulators use Windows operating systems.

Presented below, in Table 12.4.1.7.1, are a number of computers that use the UNIX Operating System, provided about seven years ago.

Table 12.4.1.8.1 List of Computers using UNIX Operating System (Year 2001)						
Computer	MIPS	Int Cache	Ext Cache	RAM	Band width CPU to CPU	Hard Drive Memory
Concurrent Power Hawk	200	64 KB	256 KB	1 GB	400 MB/Sec	1.0 GB
Concurrent Power Maxion	200	64 KB	1 MB	1 GB	400 MB/Sec	1.0 GB
SGI Origin 2000	250	32 KB	4 MB	4 GB	1.28 GB/Sec	4.0 GB
SUN Ultra Enterprise 60	?	32 KB	4 MB	2 GB	1.9 GB/Sec	4.3 GB
Intergraph TDZ 2000	?	32 KB	512 KB	256 MB	?	4.3 GB
Encore 32/67Alpha	23	32 KB	2 MB	512 MB	100 MB/Sec	4.0 GB

Presented below, in Table 12.4.1.7.2, is a computer used for the F-18 Aircraft Trainer Next Generation Threat System (NGTS) that uses the Windows Operating System.

Table 12.4.1.8.2 List of Computers using Windows 2000 Operating System (Year 2005)						
Computer	MIPS	Int Cache	Ext Cache	RAM	Band width CPU to CPU	Hard Drive Memory
2 GHz Intel XeonDual Processor	?	?	4 MB	?	21 GB/Sec	100 GB

Computer specifications change almost on a monthly basis, so the values detailed in Table 12.4.1.8.1 and Table 12.4.1.8.2 are only valid for the date indicated.

12.4.2 Flight Simulator Software

The Host Computer, illustrated in Figure 12.4.1.1, provides a large software program, usually more than a million lines of code that simulate all of the functions of a real world aircraft performance. These software modules include the following software modules

> Executive
> Instructor Operator Station
> Aural and Motion Cues
> Visual Cues
> Flight Controls
> Aerodynamics
> Propulsion
> Avionics
> Instrumentation
> Displays
> Panels, Switches, and Indicators
> Control Loader
> Visual Display System
> High Fidelity Aircraft
> Low Fidelity Aircraft
> Radar
> High Fidelity Weapons
> Low Fidelity Weapons
> Threat Tactics
> Electronic Warfare Jammers, Chaff and Flares
> Infra-red sensor

The flight simulator software program language is usually FORTRAN, C or C++, or a combination of the software languages.

Software Requirements

The typical flight simulator software requirements include the following.

1. The program software shall require compliance with SEI Capability Maturity Model (CMM) Level 3, including the coordination with the hardware, modeling, interface, human factors, security, logistics and safety.

2. The program shall support an open software system architecture, using commercial standards, non-proprietary, portable software, a standard software language and architecture for all platforms, that allows updates to be pursued.

3. The program shall support the Defense Information Infrastructure (DII) Common Operating Environment (COE). DIICOE is a software infrastructure, a collection of reusable software components, a set of application program interfaces, and a series of specifications and standards for developing interoperable systems.

4. The newly developed software or any upgrade shall feature modularity in its design. To provide ease of maintenance, modules shall be designed along functional lines, with high cohesion, and designed to minimize relationships between modules, with low coupling.

5. The computer processing speed, memory and Input (I) / Output (O) software requirements shall include 100 % spare capacity for future software growth, or a plan to provide the growth capability. The spare capability shall be measured at the highest level of processing use.

6. Software human factors must be utilized so that the man-machine interfaces shall be user-friendly, with

emphasis on simple, intuitive operation. The software shall work well, be easy to learn, interact with other products, and provide ease of software corrections and future software developments. Systems controls and displays shall emphasize logical operation by personal using equipment with minimum operator training.

7. The software on different hardware components should be different; however software re-use is encouraged, where feasible.

8. The software metric processes shall be tracked using automated software tools.

Software Design and Development

The typical flight simulator software builds include the following efforts.

1. Software Build Identification

2. Software Build requirements

3. Software Build design and Development Efforts

> Requirements Analysis
> Preliminary Design
> Detailed Design
> Peer Review
> Software Coding
> Software Implementation
> Software Integration Test
> System Integration
> Production and Development

4. Maintenance

Software Configuration Management

A typical flight simulator software configuration management process includes the following software management tools.

1. Requirements Management: IBM Rational Requisite Pro

2. Software Management: IBM Rational Clear Case

3. Change Management: IBM Rational Clear Quest

4. Test Management: IBM Rational Test Manager

Software Testing

Several different types of testing will occur. Unit testing is performed by the individual developers, and will ensure that the code modified by the developer performs correctly. The developer will also perform subsystem integration and regression testing. The developer will integrate any modifications into the most current baseline, and perform the standard integration tests. If successful, the modifications will be integrated into the new baseline.

Before a build is released, a Formal Acceptance Test (FAT) will occur. The FAT will ensure that the product meets all build requirements, and will be conducted on hardware identical to the hardware on which the final installation will occur. Informal system level testing will occur before the FAT. The Systems Test Team will perform formal Systems Integration to provide verification that the system meets its documented requirements. Details of this testing is provided in the System Test Plan (STP).

12.4.3 Flight Simulator Modeling

The Research and Trainer Flight Simulator include the following model types.

Flight Simulation Models:

Equation of Motion	Spin
Atmosphere	Carrier Bubble
Winds	Automatic Trim
Turbulence	Latitude / Longitude
Weather Systems	Magnetic Variation
IREPS	Aerodynamics
Automatic Terminal Information System (ATIS)	Weight & Balance
Carrier Control Approach (CCA)	G-Cueing
Carrier Motion	Carrier Surface Facilities
Sea State	

Aircraft Systems Models:

Air Intercept Control System (AICS)	Nose Steering
Fuel Systems	Nose Catapult
Electrical Power	Arresting Hook
Hydraulic System	Environmental Control System (ECS)
Pneumatic System	Oxygen System
Wing Sweep System	Pilot-Static System
Flap / Slats	Standby Flight Instruments
Glove Vanes	Angle of Attack (AOA) System
Speed Brakes	Canopy System
Primary Flight Controls	Ejection System
Automatic Flight Control System (AFCS)	Lighting System
Landing Gear	Control Loading System
Wheel Brake System	Control Loading Interface

Propulsion Systems Models:

Gas Generator	Engine Instruments
Bleed Air System	Engine Mode Input
Ignition System	Engine Mode Output
Starting System	Compressor Rotor Speed Demand
Oil System	Anti-Ice & Stall
Afterburner	Engine Pressure Ratio
Variable Exhaust Nozzle System	Fuel Flow
Power Generation System	Nozzle Area & Thrust
Fire Detection / Extinguishing System	Engines Electrical System
Approach Power Compensator	Engine Instruments
Engine Control Systems	Air Intercept Control System (AICS)

Communication Systems Models:

UHF Communications	Cryptic Encoder
VHF Communications	Inter-Communications System
Automatic Direction	
Finder (ADF)	Interface Blanker Unit (IBU)
IFF Transponder	

On-Board Computers Models:

Computer Air Data	
Computer (CADC)	Internal Measurement Unit (INS)
Computer System Data	
Computer (CSDC)	On Board Checkout (OBC) Interface
Armament	Weapons Flyout
Display Interface	Magnetic Tape Memory (MTM)

Tactical Displays

Navigation Systems Models:

Attitude Heading Reference	
System (AHRS)	TACAN
Integrated Logistics System (ILS)	Radar Altimeter
Data Link	Radar Beacon
Magnetic Compass	

Tactical Environment Models:

Object Dynamics	Scenario Management
Threat Tactics	Threat Aircraft Maneuver Logic
Threat Countermeasures	Relative Kinematics
Library Management	Threat Aircraft Dynamics

Sensor System Models:

Radar	Television Camera System (TCS)
IFF Interrogator	Tactical Air Reconnaissance Pod System (TARPS)

Electronic Warfare Models:

Jammers	Flares
Chaff	

Instructor Operator Station Models:

IOS Monitor	Mission Control
Mode Control	

Illustrated below, in Figure 12.4.3.1, is an example of a very detailed aircraft model block diagram. This diagram presents many of the components that simulate the threat aircraft performance. Each of the blocks in the diagram are converted initially to equations and then to software. This diagram was provided by John Morelli.

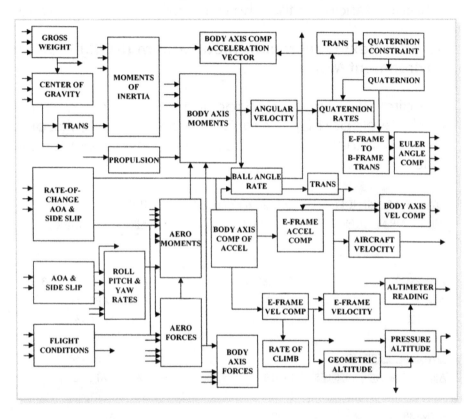

Figure 12.4.3.1 Aircraft Model Block Diagram

12.4.3.1 Trainer Flight Simulator Tactical Environment

Provided below in Section 12.4.3.2 and 12.4.3.4 are the F-18C and F-14B/A & D Trainer tactical environment flight simulator models. The two tactical environments utilized different designs so that the comparisons are interesting.

12.4.3.2 Next Generation Threat System (NGTS) Tactical Environment Models

The aircraft, weapon and radar models were designed and developed by Mesa Corporation, Mesa, Arizona The F-18 Trainer (NGTS) Tactical Environment includes the following models.

OWNSHIP & FRIENDLY AIR-TO-AIR MISSILES	THREAT & FRIENDLY AIRCRAFT	THREAT & FRIENDLY AIRCRAFT RADAR
AIM-7M/H	F/A-18C	F-18C / APG-65
AIM-9M-8	F/A-18E & F	F/A-18E & F / APG-73
AIM-9X	F-14D	F-14D / APG-71
AIM-120	F-15C	F-15C / APG-63
	F-15E	F-15E / APG-70
OWNSHIP & FRIENDLY	F-16C	F-16C / APG-68
AIR-TO-SURFACE MISSILES	EA-6B	EA-6B / APS-130
	B-1	B-1 / APQ-164
AGM-88	B-52	B-52 / APQ-156
AGM-154 (JSOW)	E-2C	E-2C / APG-125
	E-3	E-3 / APY-1/2
OWNSHIP & FRIENDLY	KC-130	KC-130 / None
BOMBS	Mirage F-1	Mirage F-1 / Cyrano IV
	Mirage 50	Mirage 50 / Aida II or Cyrano IV
MK-82	Mirage 2000	Mirage 2000 / Thompson CFI
MK-84 CFA	Mirage III	Mirage III / Cyrano II

Flight Simulator Products

GBU-31 (JDAM)	Mirage V	Mirage V / Cyrano IV
	MIG-17	MIG-17 / Spin Scan
OWNSHIP & FRIENDLY	MIG-21F-13 (Fishbed)	MIG-21F-13 / High Fix
AIR & GROUND GUNS	MIG-23ML (Flogger G)	MIG-23ML / High Lark II
	MIG-23M (Flogger B)	MIG-23M / High Lark I
570 X 20mm	MIG-25P (Foxbat A)	MIG-25P / Foxfire
270 X 30mm	MIG-25PDS (Foxbat E)	MIG-25PDS High Lark IV
HMMWV Hummer	MIG-29B (Fulcrum A)	MIG-29B / Slot Back I
M1A1 Abrams Tank	MIG-31 (Foxhound)	MIG-31 / Flash Dance
M2 Bradley IFV Tank	SU-24K (Fencer)	SU-24K / Drop Kick
M113A3 Gavin APC Tank	SU-24M (Fencer D)	SU-24M / Drop Kick
LAV-25 Tank	SU-25 (Frogfoot A)	SU-25 / None
	SU-27 (Flanker B)	SU-27 / Slot Back II
OWNSHIP & FRIENDLY	SU-27SK (Flanker B)	SU-27SK / Slot Back II
ELECTRONIC WARFARE	A-50 (Mainstay)	A-50 / A-50U
	AN-72 (Coaler)	AN-72 / None
Self-protection jamming	F-7 (Airguard)	F-7 / Type 222
Standoff jamming	F-8 (Finback)	F-8 / Type 204 ZHUK
Escort jamming	TU-22M-3 (Backfire C)	TU-22M / Down Beat
Expendable decoys	TU-95 (Bear)	TU-95 / Down Beat
Chaff	AV-8	AV-8 / APG-65
Flares	MI-24D (Hind)	MI-24D / None
	KA-25 (Hormone)	KA-25 / Big Bulge

AIRCRAFT TACTICS TECHNIQUES	AIRCRAFT JAMMING	
Pure Pursuit	Angle Gate Walk Off	Asynchronous Swept /
Notch	Barrage Noise	Wave Modulation
Drag	Spot Noise	Synchronous Swept /
Single Side Offset	Narrow Band Repeater Noise	Wave Modulation
Single Side 20 / 60	Pseudo Random Noise	Inverse Conical Scan
Bracket	Clutter Bin Masking	Repeater Swept /
Grinder	Range Bin Masking	Amplitude Modulation
Crank	Multiple False Targets	Cross Pole

Forward Quarter /
Missile Defense

Gun Defense

Skate / Leave

Range Gate Pull Off

Range / Velocity
   Gate Pull Off

Terrain Bounce

Velocity False Targets

Velocity Gate Pull Off

Linear Modulation

Missile Guidance

Scintillation

Blinking Noise

THREAT AIR-TO-AIR MISSILES

AA-2B, C & D (Atoll)
AA-7A, C, & D (Apex)
AA-8B (Aphid)

AA-9 (Amos)
AA-10A &C (Alamo)

AA-11 (Archer)
AA-12 (Adder)

THREAT AIR-TO-SURFACE MISSILES

AGM-88A

AGM-65 (Maverick)
AS-14 (Kedge)
AS-15 (Kent)
AS-16 (Kickback)
S-5K (Rocket Pod)

THREAT BOMBS

4X 220 Lb

THREAT GROUND SAM SITES

SA-2E (Guideline)
SA-3 (Goa)
SA-5 (Gammon)

& Odd Pair
SA-6 (Gainful)

 & Fire Dome
SA-7 (Grail)
SA-8 (Grecko)
SA-9 (Gaskin)
SA-10 (Grumble)

SA-12 (Giant)

SA-13 (Gopher)
SA-16 (Gimlet)
SA-18 (Grouse)
SA-19 (Grisum)
SA-20 (Triumf)

THREAT SURFACE SAM SITES

SA-N-4 (Gecko)

THREAT GROUND RADAR

SA-2 / Fan Song
SA-3 / Low Blow
SA-5 / Square Pair,
   Tall King & Odd Pair

SA-6 / Straight Flush,
   Tall King, Fire Dome

SA-7 / Grail
SA-8B / Land Roll
SA-9 / IR
SA-10 / Flapid,
   Clamshell & Big Bird
SA-12 / Grill Pan & Bill Board
SA-13 / Snap Shot
SA-16 / IR
SA-18
SA-19 / Hot Shot
SA-20

THREAT SURFACE RADAR

SA-N-4 / Pop Group

THREAT AIR-TO-AIR	SA-N-6 (Grumble)	SA-N-6 / Top Dome
GUNS	THREAT SURFACE	THREAT GROUND
60 X NR-30	SSM SITES	& SURFACE GUNS
200 X GSh-23L		
150 X Gsh-301	SS-1	2S19 Self Propelled
		Howitzer
250 X Gsh-301		T-72 MBT Tank
250 X 30mm		BMP-2 Tank
750 X 30 mm		T-80 MBT Tank
500 X Gsh-6-23M		ZSU-23-4
1000 X Gsh-23M		ZSU-57-2
1 X 23mm		
2 X 30mm		
6 X FAB-500		

SHIPS	SHIPS RADAR
Kirov CGN	Kirov CGN / Pop Group, Top Dome, Cross Sword, Front Door, Kite Screech, Bass Tilt, Top Steer, Shot Rock, Rum Tub Boghammer

12.4.3.3 Next Generation Threat System (NGTS) Tactical Environment Architecture

In 1 July 2006, the Next Generation Threat System (NGTS) Program was in Build 6 of twelve builds. The number and type of ownship and threat models will be vastly increased over the next six builds. Also, the current F-18 Trainer has an old tactical environment that is currently used, in addition to the NGTS system.

The NGTS tactical environment architecture provides a data base model, as illustrated below. A model is a set of equations with data added that simulates a system such as an aircraft, weapon or sensor. Data Based Driven models can utilize specific model equations with specific model data or generic model equations with specific model data. The Data Based Driven model that utilizes specific equations provides a higher fidelity performance than the Data Base Driven model that utilizes generic equations. The NGTS Data Base Driven model architecture utilizes generic equations. The NGTS system was initially designed to provide a medium fidelity model. The equations are converted to software to create a software model. A higher fidelity model requires a higher processing time and a larger memory requirement than a lower fidelity model. Presented below, in Figure 12.4.3.3.1, is the model data base driven architecture.

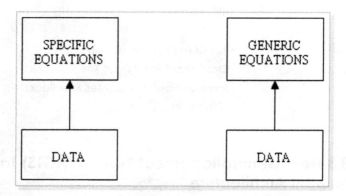

Figure 12.4.3.3.1 Data Base Driven Model Architecture

The NGTS system is designed to provide generic model equations, but with very good model data so that the final model is a fairly high performance model. The model data is obtained through a series of model data panels, which are illustrated below, in Figures 12.4.3.3. 2 through 5

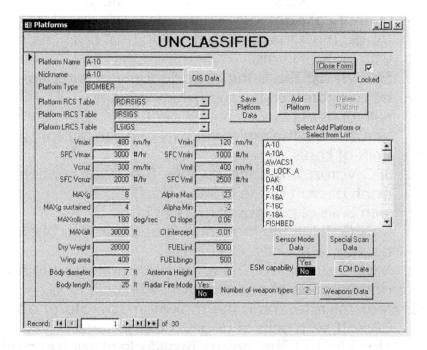

Figure 12.4.3.3.2 Aircraft Data Panel

The various aircraft data is provided in this panel for the specific Aircraft Platform identified. The aircraft data definitions are presented below.

1. Aircraft Name: This is the Aircraft name, such as F-18C.
2. Nickname: This is the aircraft nickname, such as Hornet.

3. Platform Type: This is the aircraft type, such as fighter.
4. Platform RCS Table: This is a table of the radar cross section of each aircraft (small, medium or large)
5. Platform IRCS Table: This is a table of the Infra Red intensity of each aircraft (small, medium or large)
6. Platform LRCS Table: This is a table of the Laser intensity of each aircraft (small, medium or large)
7. Vmax: This is the maximum velocity of each aircraft, with units of knots.
8. SFC Vmax: This is the specific fuel consumption of each aircraft at the maximum velocity, with units of pounds of fuel per hour.
9. Vcruz: This is the cruise velocity of each aircraft, with units of knots.
10. SFC Vcruz: This is the specific fuel consumption of each aircraft at the cruise velocity, with units of pounds of fuel per hour.
11. Max G: This is the maximum g acceleration of each aircraft, with units of G.
12. Max G sustained: This is the maximum g sustained acceleration of each aircraft, with units of G.
13. Max roll rate: This is the maximum roll rate of each aircraft, with units of degrees per second.
14. Max Alt: This is the maximum altitude of each aircraft, with units of feet.
15. Dry Weight: This is the weight of each aircraft, without fuel, with units of pounds.
16. Wing area: This is the area of each aircraft wing, with units of square feet.
17. Body diameter: This is the diameter of each aircraft body, with units of feet.
18. Body length: This is the length of aircraft body, with units of feet.
19. Vmin: This is the minimum velocity of each aircraft, with units of knots.

---

20. SFC Vmin: This is the specific fuel consumption of each aircraft at the minimum velocity, with units of pounds of fuel per hour.
21. Vmil: This is the military velocity of each aircraft, with units of knots.
22. SFC Vmil: This is the specific fuel consumption of each aircraft at the military velocity, with units of pounds of fuel per hour.
23. Alpha Max: This is the maximum angle of attack, of each aircraft, with units of degrees. The angle of attack is the angle between the wing and the direction of the wind.
24. Alpha Min: This is the minimum angle of attack of each aircraft, with units of degrees.
25. CL slope: This is the slope of the CL versus Angle of Attack curve, of each aircraft, with no units. The slope of the curve determines how rapidly CL increases with angle of attack
26. CL intercept: This is the y-intercept of the CL versus Angle of Attack curve, of each aircraft, with no units, as illustrated below, in Figure 12.4.3.3.3.

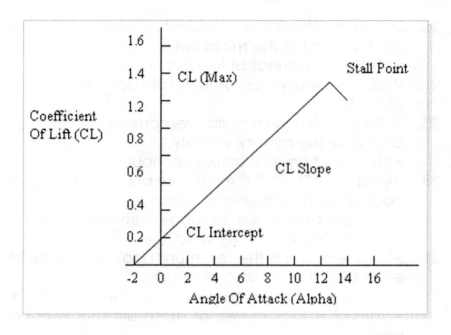

Figure 12.4.3.3.3 Coefficient of Lift versus Angle of Attack

27. Fuel Limit: This is the maximum amount of fuel load, of each aircraft, with units of pounds of fuel.
28. Fuel Bingo: This is the amount of fuel required to return to their landing base, of each aircraft, with units of pounds of fuel.
26. Antenna Height: This is the height of each aircraft antenna, with units of feet.
29. Radar Fire Mode: This indicates if the aircraft has a weapons fire mode, with units of yes or no.
30. TADIL Capability: This indicates if the aircraft has a TADIL capability, with units of yes or no.
31. ESM capability: This indicates if the aircraft has an Electronic Support Measures (ESM) capability, with units of yes or no.
32. IFF System: This identifies the specific IFF hardware.

33. Transponder / Interpolator: This indicates if the IFF hardware is a transponder or interpolator, with units of yes or no.
34. Number of weapon types: This provides the specific number of weapon types, carried by this aircraft
35. Comm. Type: This identifies the communication type, such as Mandata
36. Source Document: This provides the documents that identify the parameter values.

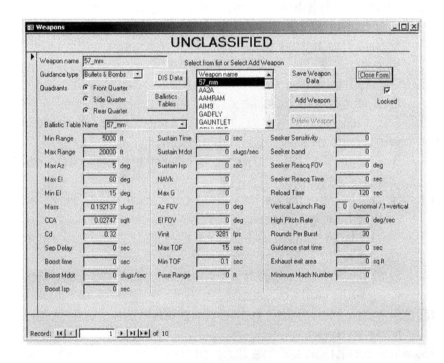

Figure 12.4.3.3.4 Weapons Data Panel

The various weapon data is provided in the panel, illustrated in Figure 12.4.3.3.4, for the Weapon Name identified. The weapon data definitions are presented below.

1. Weapon Name: This is the Weapon name, such as AA-10A
2. Guidance Type: This is the guidance type, such as Active Radar Homing
3. Min Range: This is the minimum range of the weapon, with units in feet
4. Max Range: This is the maximum range of the weapon, with units in feet
5. Max Az: This is the maximum azimuth of the weapon, with units in degrees
6. Max El: This is the maximum elevation of the weapon, with units of degrees
7. Mass: This is the mass of the weapon, with units of slugs
8. CCA: This is the CCA, with units of square feet
9. Cd: This is the coefficient of drag, with no units
10. Sep Delay: This is the time of stage separation delay, with units of seconds
11. Boost Time: This is the time of additional force, with units of seconds
12. Boost Mdot: This is the time of vapor mass flow, with units of slugs / sec
13. Boost Ist: This is the time of the first additional force stage, with units of seconds
14. Sustain Time: This is the time of sustained burning of fuel, with units of seconds
15. Sustain Mdot: This is the time sustained vapor mass flow, with units of slugs / sec
16. Sustain Ist: This is the time of the first sustained stage, with units of seconds
17. NAVk:
18. Max G: This is the maximum allowable G-loading of the weapon

19. Az FOV: This is the azimuth field of view of the weapon, with units of degrees
20. El FOV: This is the elevation field of view of the weapon, with units of degrees
21. Vinit: This is the initial velocity, with units of feet per second
22. Max TOF: This is the maximum time of flight of the weapon, with units of seconds
23. Min TOF: This is the minimum time of flight of the weapon, with units of seconds
24. Fuse Range: This is fuse range, with units of feet
25. Seeker Sensitivity: Seeker sensitivity allows greater acquisition and tracking, with no units
26. Seeker band: Bandwidth required to Hone in on wide range of radar
27. Seeker Reacq FOV: This is the seeker field-of-view, with units of degrees
28. Seeker Reacq Time: This is the seeker reaction time, with units of seconds
29. Reload Time: This is time to make a new weapon active, with units of seconds
30. Vertical Launch Flag: This is a launch flag identification, with no units
31. High Pitch rate: This is the pitch rate, with units of degrees per second
32. Rounds Per Burst: This is rounds fired per burst, with units of number
33. Guidance start time: This is the guidance time, with units of seconds
34. Exhaust exit area: This is the weapon exhaust exit area, with units of square feet
35. Minimum Mach Number: This is the weapon speed, with units of Mach

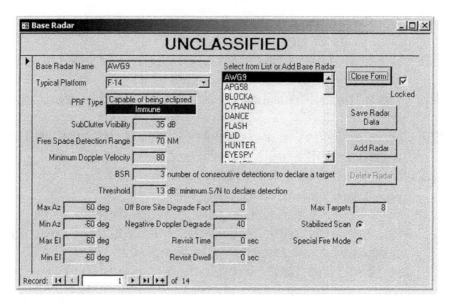

Figure 12.4.3.3.5 Radar Data Panel

The various radar data is provided in the panel, illustrated in Figure 12.4.3.3.5, for the Radar Name identified. The radar data definitions are presented below.

1. Base Radar Name: This is the Radar name, such as APG-65
2. Typical Platform: This is the aircraft that the radar is used, such as F-18C
3. PRF Type: Pulse Repetition Frequency Type is a selected term (Eclipsed, lose detection at a range) or (Immune, not lose detection at a range).
4. Sub-Clutter Visibility: This is the number of dB that the terrain around the target provides, with units of dB.
5. Free Space Detection Range: This is the radar detection range, with units of nautical miles.
6. Minimum Doppler Velocity: This is the minimum velocity of the threat aircraft that allows radar detection, with units of knots.

7. BSR: The Blip Scan Ratio is the numbers of radar scan detections required for conformation of target detection.

8. Threshold: This is the number of dB above noise and clutter that the target must provide for radar detection, with units of dB.

9. Maximum Azimuth: This is the radar right most gimbal limit, with units of degrees

10. Minimum Azimuth: This is the radar left most gimbal limit, with units of degrees.

11. Maximum Elevation: This is the radar upper gimbal limit, with units of degrees.

12. Minimum Elevation: This is the radar lower gimbal limit, with units of degrees.

13. Off Bore Sight Degrade Fact: This requires inputting a 1 or a 0 value, that either allows (1) use of an off bore site degrade table or (0) does not allow use of the off bore sight table.

14. Negative Doppler Degrade: This is the number of dB that identifies the radar sensitivity in tracking inbound targets versus outbound targets. Dividing the outbound range by the inbound range and converting the ratio into dB terms by multiplying the ratio by 10 times the base 10 log of the ratio obtain the number.

15. Revisit Time: This is the maximum time allowable for an initial radar detection to be detected again, before the detection is dropped.

16. Max Targets: This is the maximum number of targets that can be tracked simultaneously.

17. Stabilized Scan: This is a choice of "Select" or "De-select" that determines if radar will scan in relation to the horizon, as the aircraft maneuvers, or will scan in relation to the aircraft structure, regardless of its maneuvers.

18.   Special Fire Mode: This is a choice of "Select" or "De-select" that determines if radar can continue tracking the target aircraft while also tracking a weapon in flight.

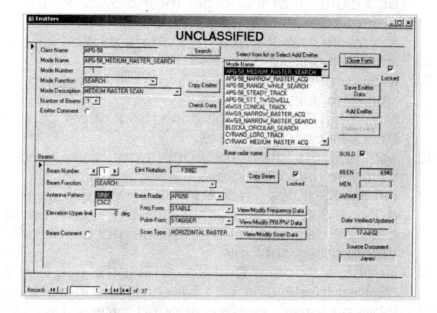

Figure 12.4.3.3.6 Emitter Data Panel

The various radar emitter data is provided in the panel, illustrated in Figure 12.4.3.3.6, for the Radar Emitter Class Name identified.

The emitter data definitions are presented below.

1.   Class Name: This is the Emitter name, such APG-65.
2.   Mode Name: This is the specific mode of the Emitter that this information is providing, such as Track While Scan (TWS) mode.
3.   Mode Number: This is a number that will be identified with the Emitter mode, such as 1 for the TWS mode.

4. Mode Function: This is a short definition to explain the Emitter mode function, such as TWS is used to track multiple targets simultaneously. (Mode choices include Search, Acquisition, Optical Search, Optical Track, Track While Scan, Track While Scan Acquisition, Track While Scan Dwell, Single Target Track, Range While Search, Fire, Optical Fire, Situational Awareness Mode, Manual Acquisition, Pseudo Track Fire, Pseudo Track Fire/Fire, Pseudo Track Fire/Medium PRF, Pseudo Track Fire/Medium PRF/Fire, Jammer, Laser Track)

5. Mode Description: This is used to provide further clarification of the Emitter mode, such as X targets can be tracked simultaneously.

6. Number of Beams: This identifies the number of beams (Beam choices include Frequency beams, PRF beams, PRI beams, Antenna Scan beams) that this emitter can provide, with a maximum of five allowed.

7. Emitter Comment: A circle is checked to open a comment window. The comments window allows the NGTS database developer to provide notes regarding the Emitter mode.

8. Beam Number: This number identifies which of the beam options this data is being provided for.

9. Beam Function: This provides a pull down menu that allows you to select the specific function of the beam whose data is being presented, such as Tracking. (Beam Function choices include Search, Height Finder, Acquisition, Tracking, Acquisition and Tracking, Command Guidance, Illumination, Range-Only-Radar, Missile Beacon, Missile Fuse, Active Radar Missile Seeker, Jammer, Beacon Interrogator)

10. Antenna Pattern: This identifies if the antenna produces a beam in a sinX or csc2 pattern.
11. Base Radar: This identifies the specific radar that the database information is provided, such as APG-65.
12. Elevation Upper Limit: This is the upper limit of the beam's elevation, with units of degrees.
13. Elevation Lower Limit: This is the lower limit of the beam's elevation, with units of degrees.
14. Beam Comment: A circle is checked to open a comment window. The comments window allows the NGTS database developer to provide notes regarding the sensor's beam.
15. ELINT Notation: This identifies the ELINT Notation utilized for NGTS, which is ELNOT.
16. Frequency: This is a pull-down menu that allows you to select an Emitter frequency domain, such as Stable. (Frequency format choices include Hopper, P2P Random, Patterned, Random List, Slider, Special, Stable)
17. Frequency Data: This is a button that opens a panel of frequency data. A yes or no is provided for FM on Pulse and Channel Count.
18. Pulse: This is a pull-down menu that allows you to select an Emitter pulse being used by the sensor, such as Stagger. (Pulse choices include Continuous Wave, Cyclic, Jitter, Random, Special, Stable, Stagger, Steady, Sync Jump, Wave Form)
19. PRI/PW Data: This is a button that opens a panel of pulse data. The Pulse Width, Pulse Repetition Interval, and Phase Delay are provided, with units of microseconds.
20. Scan Type: This is pull-down menu that allows you to select an Emitter scan, such as Horizontal Raster. (Scan Type choices include Steady, Horizontal Sector,

Horizontal Raster, Lobe on Receive Only, Conical, TWS, No, Circular).

21. Scan Data: This is a button that opens a panel of scan data. The Scan Width Azimuth, Scan Period, Number of Bars, Bar Orientation, Scan Direction Azimuth and Scan Direction Elevation are provided.

22. BEEN: This is a DIS Enumeration document number.

23. Jarm Number This is a DIS Enumeration document number.

24. Source Documents: This is a list of the documents used to provide the technical information provided in the Emitter database.

25. Comment: The comments window allows the NGTS database developer to provide notes regarding the Emitter database information.

Discussed below is the process that identifies some of the radar and emitter data terms. Let's assume a Shooter Aircraft is approaching a Target Aircraft at a range of 100 nm. The Shooter Aircraft radar is in the search mode. As the Shooter Aircraft approaches the Target Aircraft radar detection is acquired when the Shooter Aircraft radar obtains a radar signal threshold return strength of X DB, identified in the database Radar Panel. The Radar Panel Blip Scan Ratio (BSR) number of detections, identified in the Radar Panel, is used to determine if a track is found.

The Signal to Noise number (S in DB) is typically calculated from the following basic radar equation.

$$S = \frac{Pt\,(Gt)\,(Gr)\,(L)\,W^2\,(RCS)}{(4\,PI)^3\,R^4}$$

Where,

Pt is the Shooter Aircraft Radar Power (Watts), identified in the database Emitter Panel

Gt is the Shooter Aircraft Radar Transmitter Gain (DB), identified in the database Emitter Panel

Gr is the Shooter Aircraft Radar Return Loss (DB), identified in the database Emitter Panel

L is the loss over space from the Shooter Aircraft to the Target Aircraft and back (DB), identified in the database Radar Panel

W is the Wavelength (equal to 1 / Frequency, identified in the database Emitter Panel)

RCS is the radar cross-section (Sq Meters)

R is the range (Ft)

The signal to noise number is used to determine if the target will be identified by the aircraft radar computer.

12.4.3.4 F-14 B/A & D Trainers Tactical Environment Product

The F-14B/A & D Trainer tactical environment consists of high fidelity close-in-combat aircraft models, high fidelity TACTS Range weapon models and medium fidelity aircraft, weapon and radar models.

The close-In combat high fidelity aircraft models were designed and developed by McDonnell Douglas

Corporation, St Louis, Missouri. The TACTS Range high fidelity weapon models were designed and developed by First Ann Arbor Corporation (FAAC), Ann Arbor, Michigan. The medium fidelity aircraft, weapons, and radar models were designed and developed by AAI Corporation, Cockeysville, Maryland.

The F-14 B/A & D Trainers Tactical Environment includes the following models.

OWNSHIP & FRIENDLY AIR-TO-AIR MISSILES	THREAT & FRIENDLY AIRCRAFT	THREAT & FRIENDLY AIRCRAFT RADAR
AIM-7F	F-14D	F-14D / APG-71
AIM-7M	F/A-18	F/A-18 / APG-65
AIM-7M (H)	F-15	F-15C / APG-63
AIM-9H	F-16	F-16N / APG-66
AIM-9L	F-4	F-4S / AWG-10
AIM-9M	F-22	F-22 /
AIM-9X	JSF	JSF /
AIM-54C	AV-8	AV-8 / None
AIM-120A & C	E-2C	E-2C / APG-125
	EA-6B	EA-6B / APS-130
OWNSHIP & FRIENDLY AIR-TO-SURFACE MISSILES	V-22	V-22 / None
	KFIR	KFIR / EL/M-2001B
	Super Etendard	Super Etendard / Agave
AGM-65D & G (Maverick)	Tornado	Tornado / Decca 72
AGM-84	Jaguar	Jaguar / None
AGM-88A (HARM)	Mirage 2000	Mirage 200 / Thompson CFI
AGM-154 (JSOW)	Mirage F-1	Mirage F-1 / Cyrano
	MIG-17	MIG-17 / Spin Scan

OWNSHIP & FRIENDLY
BOMBS

MK-82
MK-82 (TP)
MK-82 BSU-86 (Low Drag)
MK-82 BSU (High Drag)
MK-82 (TP) BSU-86 (Low Drag)
MK-83 BSU-86 (Low Drag)

MK-83 (TS) BSU-85 (Low Drag)

MK-84

MK-76
MK-20 Rockeye

CBU-78/B
GBU-10, 12 &16 (Paveway II)
GBU-24
GBU-31 (JDAM)

MIG-21 (Fishbed)
MIG-23 (Flogger)
MIG-25 (Foxbat)

MIG-29 (Fulcrum)
MIG-31 (Foxhound)
SU-20 (Fitter)
SU-24 (Fencer)
SU-27 (Flanker)
TU-16

TU-22M (Backfire C)

TU-26

TU-95 (Bear)
TU-160

IL-38
IL-76

MIG-21 / Jaybird
MIG-23 / Highlark II
MIG-25 / Highlark IV or Foxfire
MIG-29 / Slot Back
MIG-31 / Flash Dance
SU-20 / High Fix
SU-24 / Drop Kick
SU-27 / Slot Back II
TU-16 / Puff Ball or Short Horn
TU-22M / Down Beat or Box Tail
TU-26 / Down Beat or Box Tail
TU-95 / Big Bulge
TU-160 / Down Beat or Box Tail
IL-38 / Wet Eye
IL-76 / Squash Dome

OWNSHIP & FRIENDLY
AIR & GROUND GUNS
M61-A1
GAU-8
GAU-12

OWNSHIP & FRIENDLY
ELECTRONIC WARFARE
Self-protection jamming
Standoff jamming
Escort jamming
Expendable decoys
Chaff
Flares

Flight Simulator Products

AIRCRAFT TACTICS TECHNIQUES

LONG RANGE

A-Pole
D-Cross Over
D-Lead Around
F-Pole
Lookup
Barrel Roll
Def Break
Break Away
Drag
Glib
Gun Jink
Hook
Lead Around
Pincer
Sweep

SHORT RANGE

Basic
Circling
Spiraling
Pure Pursuit
Bulls Eye
Split
Immelman
Cuban Eight
Loop
Aileron Roll
Barrel Roll
Head On
Jinking
Station Keeping
Dive
Max G Pull-Up

AIRCRAFT JAMMING

Velocity Deception, Blinked
False Targets
Velocity Deception, Cover Pulse
Velocity Deception, Doppler
Velocity deception, Sweep Frequency
Noise with Variation
VGS, Holdout and Hook
VGS, Multiple Program
Automatic Spot Noise
Multi-Spot Noise
Barrage Noise
Sweep Noise
Pulsed Noise
Intermittent Noise
Noise Swept Amplitude Modulation
Range Gate Stealers
Multiple Frequency
Random Doppler
Cross Eye
Cross Pole
Terrain Bounce (Repeater & Noise Modes)
Doppler Noise Synchronous
Doppler Noise Asynchronous
Repeater Sweep Amplitude Modulation
Chirp Gate Stealer

Flight Simulator Products

THREAT AIR-TO-AIR MISSILES	THREAT GROUND SAM SITES	THREAT GROUND RADAR
AA-2B, C & D (Atoll)	SA-2E (Guideline)	SA-2E / Fan Song
AA-6D	SA-3 (Goa)	SA-3 / Low Blow
AA-7A, C, & D (Apex)	SA-4 (Ganef)	SA-4 / Pat Hand
AA-8A & C (Aphid)	SA-5 (Gammon)	SA-5 / Square Pair, Tall King
AA-9 (Amos)	& Odd Pair	
AA-10A, B, C, D & E (Alamo)	SA-6 (Gainful)	SA-6 / Straight Flush, Tall King & Fire Dome
AA-10BR-550 (Magic)		
AA-11 (Archer)	SA-7 (Grail)	SA-7 / Grail
AA-12 (Adder)	SA-8 (Grecko)	SA-8 / Land Roll
AAAM	SA-9 (Gaskin)	SA-9 / IR
ASRAAM	SA-10 (Grumble)	SA-10 / Flap Lid, Clamshell & Big Bird
Matra R550 Magic 1		
Matra R550 Magic 2	SA-11 Gadfly)	SA-11 / Fire Dome
	SA-12A (Gladiator)	SA-12A / Grill Pan, Bill Board
THREAT AIR-TO-SURFACE MISSILES	SA-12B (Giant)	SA-12B / Grill Pan, Bill Board
	SA-13 (Gopher)	SA-13 / Snap Shot
	SA-14 (Gremlin)	SA-14 / IR
AGM-88A	SA-15	SA-15 / Scrum Half
AGM-65 (Maverick)	SA-16 (Gimlet)	SA-16 / IR
AS-14 (Kedge)	SA-19 (Grisum)	SA-19 / Hot Shot
AS-15 (Kent)	Patriot	Patriot / MPQ-53
AS-16 (Kickback)	Roland	Roland / MPOR-16
S-5K (Rocket Pod)	Crotale	Crotale / Marador II, TMD-5000
	I-Hawk	I-Hawk / Par, CWAR, HPI
	Rapier	Rapier / Blind Fire
	RB6-70	RB6-70 / Laser Guided, Giraffe
THREAT AIR-TO-AIR GUNS	THREAT SAM SITES	THREAT SURFACE RADAR
M39A2	SA-N-1	SA-N-1 / Peel Group
12.7mm	SA-N-3 (Goblet)	SA-N-3 / Head Light

Flight Simulator Products

30mm	SA-N-4 (Gecko)	SA-N-4 / Pop Group
30mm Gatling	SA-N-6	SA-N-6 / Top Dome
BK-27	SA-N-7	SA-N-7 / Front Dome
AM-23	SA-N-9	SA-N-9 / Cross Sword
DEF-A553	SM-1MR	SM-1MR /
GSH-23	SM-2MR	SM-2MR / SPG-55C
GSH-30	Sea Cat	Sea Cat / Sparrow
NR-27	Sea Gate	Sea Gate / RTN-10X
NR-30		
NR-37		
DEF-A554		

THREAT GROUND & SURFACE GUNS		THREAT SSM SITES			
SS-N-2B	CSS-N-2	2S6	1 X 114	2 X 127	23 X 406
SS-N-2C	MM-38	RBS-70	1 X 127	2 X 130	IT1 X 76
SS-N-3B	MM-40	AK-230	1 X 100	85 mm	US1 X 76
SS-N-9	OTOMAT MK	AK-630-P1	1 X 100 mm/70	100 mm	4 X 30
SS-N-12	RGM-84	AK-630	ZSU-23-4	CH2 X 25	6 X 20
SS-N-14	Sea Killer	AK-762	2 X 30	CH2 X 37	
SS-N-19	Silk Worm	SZ-1X35	2 X 35	CH2 X 57	
BGM-109		1X20	2X 57	CH2 X 130	
CSS-N-1		1X40	2 X 114	S-60	

U. S. SHIPS / SENSOR SUITE

Nimitz / MK-95, Phalanix, SPS-48, SPS-49, SPS-10-G, LN-66, SPN10

IOWA / MK-25, MK-27, MK-13, SPG-34, Phalanix, SPS-49, SPS-64, SPG-62

Ticonderoga / SPY-1A, SPQ-9, Phalanx, SPS-49, SPS-55, LN-66

Virginia / SPG-51D, SPG-55B, SPG-60, SPG-9, Phalanx, SPS-48, SPS-40, LN-66

Spruance / MK-95/115, SPG-60, SPG-9, Phalanx, SPS-40, IPDTAS, SPS-55

Charles Adams DDG / SPG51B-D, SPG-53A-F, SPS-40, SPS-10B, SPS-39

Witchita AOR / MK-95/115, Phalanix, LN-66, SPS-10B

Arleigh Burke DDG / SPY1A, Phalanx, SPS-49, SPS-55, SPS-67

Oliver Hazzard Perry / MK-92 CAS, MK-92 Stir, Phalanx, SPS-49, SPS-55

Tarawa LHA / MK-95/115, SPG-60, Phalanx, SPS-52B, SPS-40, SPS-10B, LN-66

Flight Simulator Products

FOREIGN SHIPS / SENSOR SUITE

Luda DD (Chinese) / Wasp Head, Rice Lamp, Pea Sticks, Fin Curve, Square Tie, Decca 707

Saam FFG (Iran) / Pessy, Vitex, Contraves, Flares

La Combattante II (Iran) / Has, wm-28 Oto Melara Compact, M-20, Decca 1226, Decca 1229

Leander Class (India) / None

Najin (North Korea) / Drum Tilt, Pot Drum, Slim Net

Generic Patrol boat / None

Wadi M'Ragh (Libya) / RAN 11/12, RTN-10X

RUSSIAN SHIPS / SENSOR SUITE

Tbillisi CVHG / Sky Watch, Plate Steer, Cake Stand

Kiev (Baku) CVHG / Cross Sward, Pop Group, Head Light, Front Door, Kite Screech, Brass Tilt, Plate Steer, Rum Tub

Kiev (Other) CVHG / Pop Sword, Head Light, Front Door, Owl Screech, Brass Tilt, Top Steer, Strut Pair, Palm Front, Don Kay

Kirov CGN / Pop Group, Top Dome, Cross Sword, Front Door, Kite Screech, Brass Tilt, Top Steer, Shot Rock, Rum Tub

Slava CG / Pop Group, Top Dome, Front Door, Kite Screech, Brass Tilt, Top Pair, Top Steer, Palm Frond, Shot Rock

Sovremenny DDG / Front Dome, Band Stand, Kite Screech, Brass Tilt, Top Steer, Palm Frond, Shot Rock

Udaloy DDG / Cross Sword, Eye Bowl, Kite Screech, Brass Tilt, Top Plate, Strut Par, Palm Frond, Shot Rock

Nanuchka II PGG / Pop Group, Band Stand, Band Stand, Muff Cobb, Peel Pair, Mius, Shot Dome

Krivak II FFG / Pop Group, Eye Bowl, Owl Screech, Head Net, Don II, Don Kay, Low Trough, Palm Frond, Shot Rock

Kresta II CG / Peel Group, Scoop Pair, Muff Cobb, Brass Tilt, Big Net, Don Kay, Don II, Palm Frond, Shot Rock

Mod Kashin DDG / Peel Group, Plank Sheave, Owl Screech, Brass Tilt, Head Net, Big Net, Don Kay, Don II, Palm Frond, Shot Rock

Flight Simulator Products

Kara CG / Pop Group, head Light, Top Bowl, Eye Bowl, Owl Screech, Brass Tilt, Head
     Net, Don Kay, Don II, Palm Frond, Shot Rock, Top Sail
Osa PTG / Drum Tilt, Square Tie
Grisha FFL / Pop Group, Muff Cob, Brass Tilt, Strut Curve, Don II
Tarantul PGG / Plank Sheave, Brass Tilt, Mius
Echo II SSGN / Front Door, Snoop Tray, Snoop Slab
Mike SSN / Snoop Head
Foxtrot SSN / Snoop Tray, Snoop Plate
Generic Mership / Don Kay, LN-66
Boris Chilikin AOR / Don Kay
Brezhnev / Plate Steer, Cake Stand
OSA II PTG / Drum Tilt, Square Tie
Charlie II SSGN / None
Oscar SSGN / None

12.4.3.5 F-14 B/A & D Trainers Tactical Environment Architecture

The F-14B/A & D Trainer tactical environment provides three different model architectures. The Close-In-Combat aircraft models utilize data base driven models, with specific equations, as illustrated below, in Figure 12.4.3.5.1. A model is a set of equations with data added that simulates a system such as an aircraft, weapon or sensor. The equations are converted to software to create a software model. The Data Based Driven model that utilizes specific equations provides a higher fidelity performance than the Data Base Driven model that utilizes generic equations. The F-14B/A & D Trainer data is hard coded in the data base driven models.

The high fidelity missile models, utilize self-contained model architecture, as illustrated below, in Figure 12.4.3.5.1. These models are the same models that are utilized in the Tactical Air Combat Trainer (TACTS Range) system. The data is hard coded in the model.

The low fidelity aircraft and missile models utilize data base models, with generic equations, as illustrated below, in Figure 12.4.3.5.1 The data for these models is also hard coded in the model. These models provide a much lower computer processing time and memory requirement, than the high fidelity models.

The F-14B/A & D Trainer system was designed to provide high fidelity models to conduct Air Combat Maneuvering (Dog fight) mission scenarios and low fidelity models to conduct many other mission scenarios.

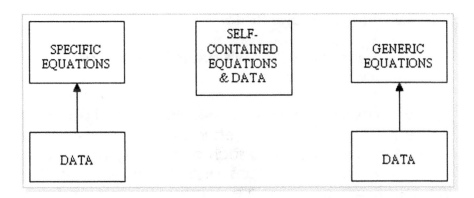

Figure 12.4.3.5.1 Data Base Driven and Self-Contained Model Architecture Illustration

12.4.3.6 Flight Simulator Radar

From the website http://en.wikipedia.org/wiki/radrar, the following radar information is provided.

Radar is a system that uses electromagnetic waves to identify the location, direction and speed of both moving and fixed objects such as aircraft, ships, motor vehicles, weather formations, and terrain. A transmitter emits radio waves, which are reflected by the target and detected by a receiver, typically in the same location as the transmitter.

The use of radio waves to detect the presence of distant objects using radio waves was first implemented in 1904, by Christian Hulsmeyer. Christian demonstrated the feasibility of detecting the presence of a ship in dense fog, but not its distance.

World War II precipitated the research to find better resolution, more portability, and more radar features.

The amount of power Pr returning to the receiving antenna is given by the radar equation presented below.

$$Pr = (Pt)(Gt)(Ar)(Q)(F)^4 \ / \ (4Pi)^2(Rt)^2(Rr)^2$$

Where,

Pt = transmitter power
Gt = gain of the transmitting antenna
Ar = effective aperture of the receiving antenna
Q = radar cross section
F = pattern propagation factor
Rt = distance from the transmitter to the target
Rr = distance from the target to the receiver

One way to measure the distance to an object is to transmit a short pulse of radio signal and measure the time it takes for the reflection to return. Since radio waves travel at the speed of light (186,000 miles per second) accurate distance measurement requires high performance electronics.

Another form of distance measuring radar is based on frequency modulation. Frequency comparison between two signals is considerably more accurate than timing the signal. By changing the frequency of the returned signal and comparing that with the original, the difference can be easily measured.

A radar has the following components,

Transmitter: A transmitter generates the radio signal with an oscillator and controls its duration by a modulator.

Antenna: An antenna broadcasts a radio signal which will spread out in all directions, and will receive signals equally from all directions.

Waveguide: A wavelength links the transmitter with the antenna.

Duplexer: The duplexer serves as a switch between the antenna and the transmitter and the receiver.

Receiver: The receiver detects the return radio signal.

Electronic System: The electronic system controls all the radar devices, to perform the radar scan, processes the return signals, and presents them on a display

Presented below, in Figure 12.4.3.6.1, is a typical ownship radar operation. A threat aircraft flies by an ownship radar system. The aircraft radar display provides the indication of the threat aircraft, first in the clear area and then in the clutter area.

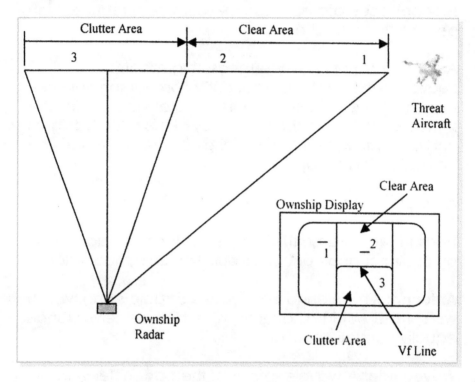

Figure 12.4.3.6.1 Typical Ownship Radar Operation

Presented in Figure 12.4.3.6.2, are a Pulse Doppler Search radar display and a Pulse Search radar display. The threat aircraft location is presented on the radar display. These displays are copied from the F-14 Aircraft Natops Manual.

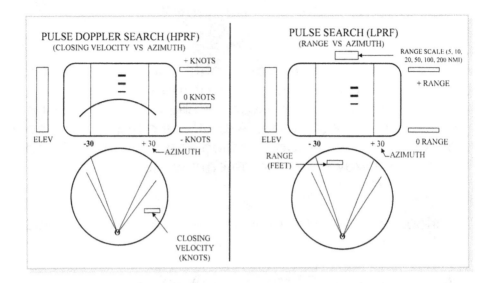

Figure 12.4.3.6.2 Pulse Doppler Search Radar and Pulse Search Radar Presentations

There are many types of aircraft radar, including the following.

Fire Control Radar: Fire Control Radar provides target azimuth, elevation, range and velocity. This radar typically emit narrow, intense beam of ratio waves to ensure accurate tracking information and to minimize the chance of losing track of the target.

Pulse Radar: Pulse Radar provides radio waves, emitted in discrete pulses. Pulse radar operates on different frequencies for different purposes.

Tracking Radar: Tracking Radar is designed to accurately track the target in range, velocity and bearing. Automatic detection and tracking radar consists of targets that show up on a radar screen as tracks, rather than as discrete blips.

Phased Array Radar: Phased Array Radar track many targets at the same time.

Continuous Wave Radar: Continuous Wave Radar transmits and receives signals at the same time. The radar can distinguish the weak returning signals from the strong transmitted signals. Aircraft use a particular type of continuous wave radar; called frequency modulated continuous wave radar, to determine their height above the ground.

Various radar modes include the following presented below, are taken from the website

Velocity Search (VS) Mode: The Velocity Search radar mode provides maximum detection range capability against a target directly in front of the ownship radar.

Range-While-Search (RWS) Mode: The Range-While-Search radar mode provides detection range capability against all targets at all aspect angles.

Track-While-Scan (TWS) Mode: The Track-While-Scan radar mode provides a combination of automatic detection and tracking functions within search radar. The primary function is to maintain target tracks in a computer with periodic information updates from a scanning radar.

Single-Target-Track (STT) Mode: The Single-Target-Track radar mode observes only one target at a time with a small azimuth angle, and continually illuminates the target with electromagnetic energy.

Presented below are a number of radar terms.

Constant False Alarm Rate (CFAR): CFAR is a method relying on the fact that clutter returns far outnumber echoes from targets of interest.

Clutter: Clutter refers to actual radio frequency (RF) echoes returned from targets which are by definition uninteresting to the radar operators in general. Such targets mostly include natural objects, such as ground, sea, precipitation, sand storms, animals, atmospheric turbulence, and buildings.

Interference: Radar systems must overcome several different sources of unwanted signals in order to focus only on the actual targets of interest.

Jamming: Radar jamming refers to radio frequency (RF) signals originating from sources outside the radar, transmitting in the radar's frequency and thereby masking targets of interest.

Noise: Signal noise is an internal source of random variations in the signal, which is inherently generated to some degree by all electronic components. Noise typically appears as random variations superimposed on the desired echo signal received in the radar receiver.

Polarization: In the transmitted radar signal, the electric field is perpendicular to the direction of propagation, and this direction of the electric field is the polarization of the wave. Radars use horizontal, vertical, linear and circular polarization to detect different types of reflections.

Presented below, in Table 12.4.3.6.1, are the radar frequency bands.

Table 12.4.3.6.1 Radar Frequency Bands		
Band Name	**Frequency Range**	**Wavelength Range**
HF	3 to 30 MHz	10 to 100m
P	Less than 300 MHz	1 m
VHF	50 to 330 MHz	.9 to 6 m
UHF	300 to 1000 MHz	.3 to 1 m
L	1 to 2 GHz	15 to 30 cm
S	2 to 4 GHz	7.5 to 15 cm
C	4 to 8 GHz	3.75 to 7.5 cm
X	8 to 12 GHz	2.5 to 3.75 cm
Ku	12 to 18 GHz	1.67 to 2.5 cm
K	18 to 27 GHz	1.11 to 1.67 cm
Ka	27 to 40 GHz	.75 to 1.11 cm
mm	40 to 300 GHz	1 to 7.5 mm
Q	40 to 60 GHz	5 to 7.5 mm
V	50 to 75 GHz	4 to 6 mm
E	60 to 90 GHz	3.33 to 6 mm
W	75 to 110 GHz	2.7 to 4 mm

12.4.3.7 Flight Simulator Jammer Model

Jamming degrades a threat radar detection range, closing velocity, azimuth and elevation detection. These jammers are called Deceptive Jammers. In addition, noise is also sometimes provided to degrade threat radar. These jammers are called Noise Jammers. Standoff jammers are usually noise jammers and self-protection jammers are usually deception jammers. Presented below, in Figure 12.4.3.7.1, is an illustration of an aircraft track flying through an enemy radar field without radar jamming and an aircraft track flying through an enemy radar field with radar jamming.

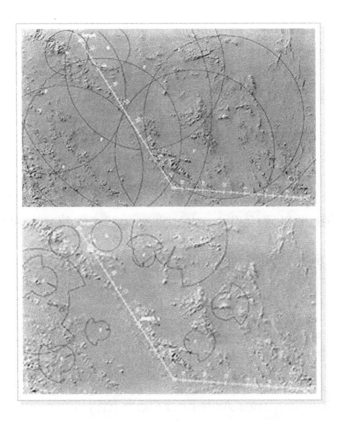

Figure 12.4.3.7.1 Aircraft Track through Enemy Radar Field with and without Jamming

A range jammer system basically operates in the following way. The ownship radar transmitter sends out a radio signal that bounces off a threat aircraft and returns to the ownship aircraft receiver. The Threat Aircraft detects the ownship radar signal, and after a delay, sends a stronger signal back to the ownship computer. The ownship radar computer thinks that the stronger signal is the actual radar return signal.

Presented below, in Figure 12.4.3.7.2, is a typical radar jammer circuit, obtained from a Pt MUGU Naval Base Jammer Report.

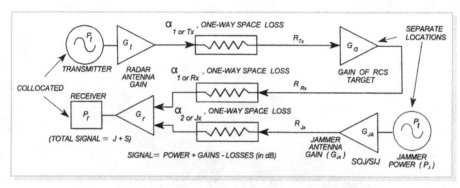

Figure 12.4.3.7.2 Typical Radar Jammer Circuit

Presented below is the Jamming-to-Signal (J/S) equation.

$$J/S = (Pj)(Gj)(4\ Pi)(Rt^4)(BWr) \ / \ (Pt)(Gt)(Q)(Rj^2)(BWj)$$

Where,

Pj = Jammer Power (dB)

Pt = Radar Power (dB)

Gj = Jammer Antenna Gain (dB)

Gt = Radar Antenna Gain (dB)

Rt = Range from Radar to Target (ft)

Rj = Range from Jammer to Radar (ft)

Q = Target Radar Cross Section (ft^2)

BWr = Radar Band Width (watts / mHz)

BWj = Jammer Band Width (watts / mHz)

The Jammer Signal (J) / Radar Signal (S) provide the ratio of jammer effectiveness. If the jammer return signal is greater than the radar return signal then the radar is jammed. If the radar signal is greater than the jammer signal then the radar is not jammed. The later situation is called burn through.

Presented below, in Figure 12.4.3.7.3, is the Jammer Control Panel.

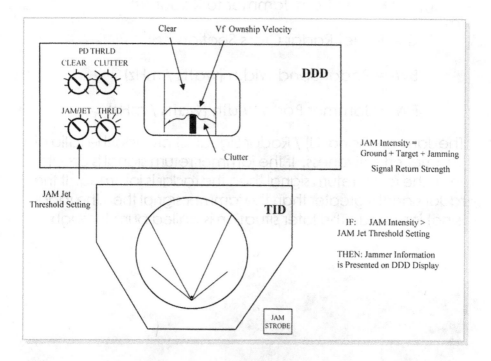

Figure 12.4.3.7.3 Jammer Control Panel

The aircraft Jammer Control Panel is used to adjust the sensitivity and threshold of the threat jammer signals. This jammer control panel is copied from the F-14 Aircraft Natops manual. Sensitivity is the susceptibility of the system to receive any signal including a jammer signal. The higher the sensitivity the better chance you will receive a jammer signal, even a weak one. The threshold is adjusted to limit the detection of all incoming signals.

The jammer types, provided in Table 12.4.3.7.1 below, are included in the F-18 and F-14 Aircraft Trainer.

Table 12.4.3.7.1 F-18 and F-14 Trainer Jammer Techniques	
Degradation Type	**Jammer Technique**
Noise	Noise with Variation Automatic Spot Noise Multi-Spot Noise Barrage Noise Sweep Noise Pulsed Noise Intermittent Noise Noise Swept Amplitude Modulation Doppler Noise Synchronous Doppler Noise Asynchronous Multiple Frequency Random Doppler
Range	Range Gate Stealers
Closing Velocity	Velocity Deception, Blinked Velocity Deception, Cover Pulse Velocity Deception, Doppler Velocity deception, Sweep Frequency VGS, Holdout and Hook VGS, Multiple Program Chirp Gate Stealer
Azimuth and Elevation	Cross Eye Cross Pole
General	False Targets Terrain Bounce (Repeater & Noise Modes) Repeater Sweep Amplitude Modulation

The F-18 and F-14 Aircraft Trainer Fleet Project Team have requested that the following jammer techniques should be developed and training exercises provided. They also requested that a number of classified foreign jammer systems be developed and training exercises provided.

Incoherent Oscillations
Gate Stealing / False Doppler Targets
Terrain Bounce
Cooperative Blinking
Cooperative Soft Blinking
Non-Adaptive Cross Pole
Adaptive Cross Pole
Digital RF Memory

Presented below are a number of jammer terms.

Angle Jamming: Any deceptive jamming specifically intended to degrade angle tracking by the victim receiving system.

Angle Deception Jamming: Any jamming technique intended to deceive a victim by posing false azimuth and elevation information. Normally, angle deception jamming is employed against a target tracking radar or missile system to inhibit or to break angle track on a true target.

Barrage Jamming: Any continuous radio frequency (RF) jamming energy distribution, whether generated from a single or from multiple jammer carriers, whose bandwidth is much greater than the bandwidth of its intended victim.

For radar jamming applications, bandwidths of 20 MHz or greater are considered barrage.

Barrage Noise Jamming: Any barrage jamming from a single or multiple noise modulated jamming carriers.

Blanking: The temporary brief keying off of a jamming signal for a purpose such as look through.

Blinking: A form of low rate on / off keying of individual jammer signals performed by multiple simultaneous aircraft for the purpose of angle deception of radar guided missiles.

Cooperative Jamming: Any jamming technique that employs two or more separate jamming platforms against a common victim.

Cooperative Blinking: A form of cooperative jamming in which two or more closely spaced aircraft key their jamming signals on and off interdependently for the purpose of deceiving, in single, a radar tracker or missile seeker.

Cross Polizarization Jamming: A jamming technique in which the jammer transmits a polarization orthogonal to that of the victim receiver. Cross polarization jamming can be adaptive or nonadaptive. Adaptive is cross pol jamming in which the jammer system determines in the real time the appropriate polarization to emit by determining the polarization it receives from the victim emitter. Non-adaptive is cross pol jamming in which the transmit polarization is predetermined from prior information.

Cover Pulse Jamming: Any active jamming technique in which the jammer attempts to produce a pulse that overlies in time, a target skin paint pulse.

Deceptive Jamming: Any jamming intended specifically to deceive any element in the receiving system.

Escort Jamming: Jamming performed as protection of other aircraft by a jamming platform flying with the aircraft that it is protecting.

False Target: Any jamming phenomenon that appears in some or all respects as a true target to a receiving system element. The false targets need not have a deceptive jamming purpose.

Home on Jam (HOJ): HOJ has the ability of some radar / missile systems to develop tracking information from the angular coordinates of a jamming emitter, usually a noise jammer.

Inverse Gain Jamming: An angle deception jamming technique involving an adaptive amplitude modulation at or near the scan rate of an active scanning radar tracker.

Jammer: The physical source of jamming, including elements such as transmitting equipment, antennas, control systems, associated receiving systems.

Jamming: Any means by which any form of interference is intentionally produced to cause degradation of the operation of any electromagnetic receiving system.

Look Through: Any change of state in an active jamming emitter that is accomplished intentionally to allow associated receiving elements to observe illuminating transmitter systems.

Noise: Any characteristics varying in a random manner.

Passive Jamming: Any means by which interference is produced intentionally without active radiation.

Pulse Up Jamming: Any pulsed jamming in which the reduced duty factor of the jammer transmitter is used to increase its instantaneous peak output power.

Random: Any variation in a jamming characteristic which occurs with approximately equal probability at any frequency or at any interval within arbitrarily established limits.

Range Gate Pull Off (RGPO): The standard active jamming technique used in range gate stealer whereby the jamming system creates a sequential programmed delay in the repeater pulse train in order to deceive a range tracking receiver element.

Range Gate Stealer: A specific type of deceptive jamming designed to cause a range tracking receiving system to interpret false data as true.

Responsive Jamming: Any active jamming which exhibits changes of state as a direct result of the receipt of information from transmitter system whose associated receiving system is a jammed target.

Self Protection Jamming: Jamming performed by an aircraft to protect itself.

Spot Noise Jamming: Any noise jamming technique whose RF bandwidth is relatively narrow compared to the bandwidth of its intended victim.

Standoff Jamming: Any jamming accomplished by a jamming platform specifically to protect other aircraft that are closer to the victim than the jamming platform.

Transponder: As applied to jamming a particular type of jammed that operates by producing noise bursts or pulses in response to arriving radar pulses.

Velocity Gate Pull Off (VGPO): Velocity gate pull off is the standard technique used by velocity gate stealers whereby the jammed creates a sequentially programmed false Doppler velocity in a velocity tracking receiving element.

Velocity Gate Stealer: Any jamming intended to deceive a Doppler receiving system by causing the victim to interpret false velocity as true.

12.4.3.8 Infrared Search and Track Sensor

The Infrared Search and Track Sensor is a passive long-wave infrared sensor system that searches for and detects heat sources within a field-of-view. The system provides the aircraft mission computer track file data on all targets while simultaneously providing infrared imagery to the cockpit display. The Infrared Search and Track Sensor provide the aircraft crew situational awareness while enhancing the engagement range of modern high performance weapons.

The F-14D Trainer model is detailed in a report entitled, "AN/AAS-42 Infrared Search and Track (IRST) System Detection Model", Naval Air Warfare Center, Warminster, Pennsylvania.

The Infrared Search and Track Sensor model uses a set of conditions describing the target and the environment, and calculates the signal-to-clutter ratio and the probability of detection to generate tracks for presentation to the IRST operator.

The following model inputs are required,

1. Location and season, selected from a set of six situations, including a. tropical, b. mid-latitude summer, C. mid-latitude winter, d. sub-arctic summer, e. sub-arctic winter, and f. United States Standard.

2. Target Type: Selected from a list of 40 types.

3. Clutter Case: Selected from a list of 24 combinations of sky and surface clutter conditions.

4. Target Heading, Bearing and Speed: Continuously variable.

5. Sensor and Target Altitude: Continuously variable.

6. Separation Range: Continuously variable.

The following Infrared Search and Track Sensor component models are required.

1. Elevation Angle Module: This module is used to calculate the elevation angle of the target, and of the cloud   horizon. A lookup table is provide with the environment conditions above the horizon (clear sky or cloudy sky)and the environmental conditions below the horizon (calm sea, rough sea, terrain, mountains, city, smooth clouds, low clouds, medium clouds, high clouds, and very high clouds).

2. Air Temperature Module: This module is used to calculate air temperature for a given target altitude and model atmosphere.

3. Skin Temperature Module: This module is used to calculate target skin temperature for a given air temperature and target speed.

4. Skin Radiance Module: This module is used to calculate the target skin radiance for a given skin temperature.

5. Target radiance Module: This module is used to determine the transmission between the target and sensor for a given target scenario. Values are pre-computed using atmospheres at eight sensor

altitudes, 13 targets altitudes and five ranges. For each scenario, six transmissions are given, corresponding to the standard atmospheres.

6. Path Radiance Module: This module is used to determine the path radiance for a given target scenario. Values are pre-computed using atmospheres at eight altitudes, 13 targets and five ranges. For each scenario, six background radiances are given, corresponding to standard atmospheres.

7. Background Radiance Module: This module is used to determine the background radiance for a given sensor altitude and target elevation angle. Values are pre-computerd using atmospheres at eight altitudes. The elevation angles arise from the combinations of sensor altitude, target altitude, and range used to determine the path radiance and target transmission tables.

8. Contrast Radiance Module: This module is used to calculate the apparent contrast radiance between the target and background.

9. Aspect Angle Module: This module is used to calculate the aspect angle of the target on the relative positions of the aircraft.

10. Target Angle Module: This module is used to calculate the projected area of the target based on the target type and aspect angle of the aircraft. The target is approximated by a four-faceted solid. The length, width and height of the target is determined from a table, which was derived so that a three orthogonal projections of the model (top, front and side) match

the projected areas of the corresponding views on actual aircraft.

11. Contract Irradiance Module: This module is used to calculate the contrast irradiance between the target and background.

12. Relative Minimum Detectable Target Irradiance (MDTI) Module: This module is used to calculate the minimum detectable target irradiance for given clutter and target conditions express relative to the MDTI under ideal conditions. The effective MDTI is based on the relative clutter and a factor which accounts for the extended target effects. Given the target projected area, range, relative clutter level and clutter state, the contrast reduction factor is calculated.

13. Signal-to-Clutter Ratio (SCR) Module: This module is used to calculate the signal-to-clutter ratio, using a number of software flags.

14. Detection Probability Module: This module is used to calculate the single-frame detection probability as a function of signal-to-clutter ratio. The detection probability is determined from a pre-computed table, which includes signal-to-clutter values translated to detection probability values.

Presented in Table 12.4.3.8.1, are the F-14D Trainer Infra-Red Sensor, Video Scan Converter and Forward Looking Infra-Red (FLIR) system specifications and display presentations.

Table 12.4.3.8.1 F-14D Trainer Infra-Red Sensor, Video Scan Converter and Forward Looking Infra-Red (FLIR) System

Specification	Infra-Red Sensor Model	IR Video Scan Converter	FLIR
Field-of-View	150 deg Horizontal by 20 deg Vertical	150 deg Horizontal by 20 deg Vertical	1-1/2 degree Horizontal by 2 deg Vertical & 6 deg Horizontal by 8 deg Vertical
Targets	10 Air-to-Air 10 Air-to Ground	View of targets in field-of-view	View of targets in field-of-view
Modes	TWSM, TWSA, Hot IR, STT, PAL, PLM	One, Two and Four Bar Scan	Two narrow field-of-views
Special Features	Detects targets in wide field-of-view, processes target information and tracks target.	Provides low resolution view of targets over large field-of-view	Provides high resolution view of targets over small field-of-view & provides target track
Background	Clear sky, Cirrus sky, Calm sea, Rough sea, Smooth cloud deck, Rough cloud deck, Desert, Woods, Mountain and Urban		
Trainer Equipment Required	Encore 32-67 Computer	Silicon Graphics Onyx Computer with Ethernet connection.	Update of video scan converter Silicon Graphics Onyx Computer
Crewstation Presentation			

12.4.3.9 Mult-Sensor Correlation

Multi-Sensor Correlation or data fusion is defined as the process of coordinating data from aircraft sensors, such as radar and infrared search and track sensors, to provide the best common situation awareness knowledge of the target.

The following is extracted from a report entitled, "F-14D Multi-Sensor Correlation Program" by the Naval Air System Command.

Multi-sensor Correlation is the combination of data from multiple sources to provide a complete, timely, tactical picture. Generically, data fusion provides the means for coordinating sensor usage, interpreting data, displaying integrated results and assessing situations. When applied to an existing tactical aircraft, the term takes the form of an avionic suite coordination enhanced by data correlation.

The concept consists of three elements: a. the spatial correlation of track from the ownship sensors, b. the classification of a track as either friend, hostile or unknown or c. the identification of the platforms with an attached confidence level.

Individual track files from each sensor serve as inputs to the multi-sensor correlation process. This data set would include: range, range rate, azimuth, elevation, direction cosines, ownship position, heading, roll, pitch, yaw, altitude, the velocity and acceleration terms for each target track, plus individual sensor identification data.

The first step in the correlation process will be spatially correlate the information from each source and develop a merged system track file. Next, the merged track data is compared to the stored data base to classify and identify the target. After the tracks have been classified and identified this data can be used to prioritize the attack and allocate weapons to the target.

Artificial intelligence can be used to provide the methodology for developing a multi-sensor correlation system.

12.4.4 Flight Simulator Interface

The Host Computer interfaces with the Aircraft Flight Simulator are illustrated in Figure 12.4.1.1

The interfaces include the following component signals.

Crewstation and Instructor Operator Station: Cockpit Flight Controls, Instruments, Switches and Lights Interface

1. Analog / Digital (A/D) signals simulate the aircraft flight controls by providing a servo-mechanism potentiometer signal between - 10 volts to + 10 volts. As the Flight Simulator simulated flight controls (stick, throttle and pedals) are rotated the − 10 to + 10 volt signal is detected.

2. Digital / Analog (D/A) signals simulate the aircraft instrumentation by providing a servo mechanism potentiometer signal between − 10 volts to + 10 volts. The Host Computer software program determines the proper flight simulator instrument position. A − 10 to + 10 volt digital signal is converted to an analog signal and the servo-mechanism rotates the instrument readout to the proper position.

3. Indiscrete signals simulate switch closures by providing a ground or open signal as the flight simulator switch is moved from one position to another.

4. Outdiscrete signals simulate lights by providing a power or open signal to turn on a flight simulator light.

The Host Computer versus Cockpit Crewstation Interface includes a J-Box Interface. The J-Box Interface is a non-computer software interface which simply distributes the crewstation power signals (120vac, 28vdc, +10vdc, -10vdc), power grounds, A/D signals, D/A signals, Indiscrete signals, and Outdiscrete signals. The J-Box Interface can be very useful in checking out the flight simulator system. Careful distribution of the power grounds is extremely important, because many fight simulator electrical problems can be traced back to having ground loops.

Crewstation and Instructor Operator Station: Displays Interface

The Host Computer provides display image information to an image generator system. The image generator system in turn provides the cockpit display and real world visual display system presentation. A basic Flight Simulator Display Interface system is presented below in Figure 12.4.4.1, extracted from a report entitled "Implementation of Hardware for Image Display and Processing" by North Carolina State University, 1999.

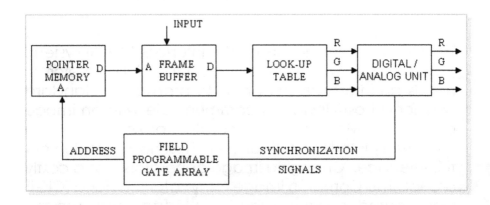

Figure 12.4.4.1 Flight Simulator Display Interface System

Field Programmable Gate Array (FPGA): The FPGA is the controller for the system address generator, sync generator; vital mode registers, and address decoder. The controller generates the enable signals for the tri-state buffers, flip flops and memories. The address generator generates sequential addresses for raster scan. The sync generator generates "blank" and "sync" signals for the system. The video mode registers determine the video mode and are written by the Peripheral Component Interface (PCI) interface. The address decoder decodes the address created by the PCI interface to assert the enable signals.

Pointer Memory: The Pointer Memory provides the frames to be scanned in an orderly fashion.

Frame Buffer: The Frame Buffer provides the storage of the image frames.

Look-Up Table: The Look-Up Table is the memory unit for color mapping.

Digital to Anal log (D / A) Unit: The D / A Unit converts the red, green, blue pixels to an analog signal to the display system.

The Common Image Generator Interface (CIGI) system, presented below from the website http://cigi.sourceforge. net is an interface designed to promote a standard way for a host device to communicate with an image generator. Most image generators provide their own proprietary interface so that every host had to implement this interface. Changing image generators was a costly ordeal. The Common Image Generator Interface (CIGI) was created to standardize the interface between the host and the image generator so that little modification would be needed to switch image generators. The CIGI system user group includes The Boeing Company and Naval Training Center, Orlando Florida.

Flight Simulator to Flight Simulator Interface

The Flight Simulator to Flight Simulator interface system is usually provided with the following systems.

1. Distributive Interactive Simulation (DIS)

The Distributed Interactive Simulation (DIS) is a standard for conducting real-time platform level wargaming across multiple host computers.

A Protocol Data Units (PDU) is transmitted across the DIS network. The DIS interface will transmit and receive DIS PDUs from the application to the network. This entails bundling and unbundling of the PDUs. Bundling is the ability to place more than one DIS PDU in a network data packet to increase network efficiency. The application needs to decode and encode the PDU based on the current DIS standard being used. To reduce the amount of network traffic, dead reckoning is used to estimate the position and orientation of an entity. The DIS Interface Functional Requirements Document describes what is required of a DIS interface.

2. High Level Architecture (HLA)

The High Level Architecture (HLA) is a general purpose architecture for distributed computer simulation systems to allow one computer simulations to communicate with other computer simulations. Using HLA, computer simulations can communicate to other computer simulations regardless of the computing platforms.

The High Level Architecture (HLA) consists of the following components.

(1) Runtime Infrastructure (RTI): The Runtime Infrastructure (RTI) provides communication between simulations. The RTI provides a programming library and an application programming interface (API) compliant to the interface specification.

(2) Object Model Template (OMT): The Object Model Template (OMT) specifies what information is communicated between simulations and how it is documented.

(3) Federate: A federate is a HLA compliant simulation.

(4) Federation: A Federation has multiple simulations connected by the Runtime Infrastructure (RTI) using an Object Model Template (OMT).

(5) Federation Object Models (FOM): The Federation Object Models (FOM) is a library of simulation components.

(6) Object: An Object is a collection of related data sent between simulations. Objects have attributes (data fields).

(7) Interactions: Interactions are events between simulations. Interactions have parameters (data fields).

(8) HLA Rules: HLA Rules are simulation rules that must be obeyed to be compliant to the standards.

The interaction of the Next Generation Threat System (NGTS) tactical environment with external flight simulators is both a DIS Interface and a HLA Interface. The NGTS interface includes a DIS / HLA Logger, which provides flight simulator parameters during a training exercise. Unfortunely the DIS / HLA Logger is not able to provide interactive data, such as range, aspect angle or closing rate, which involve two aircrafts of the training exercise. These values can be calculated from parameters that can be obtained from the DIS / HLA Logger.

Presented below, in Figure 12.4.4.2, is the Next Generation Threat System (NGTS) Block Diagram, which presents the HLA / DIS Interface. The NGTS components include the core NGTS software, the Instructor Operator System software, a separate system for high fidelity ownship weapons and an interface with a Trainer Flight Simulator. The NGTS system is utilized on the F-18 Aircraft Trainer.

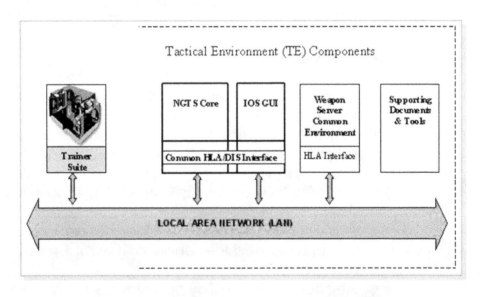

Figure 12.4.4.2 Next Generation Threat System (NGTS) Block Diagram

A number of NGTS units can be linked together to provide additional tactical environment capability. A weapon can be fired from one NGTS unit and hit an aircraft in another NGTS unit.

12.4.5 Flight Simulator Human Factors

The Systems Engineering human factors element in the design and development of a Research or Trainer Flight Simulator is extremely important. The main goal for the design and development of a Trainer Flight Simulator is to ensure that the training capability will have no negative training effects during the exercise of each training mission.

Negative Training is defined to be any element of the training system that does not present itself as it would in the real world within a defined tolerance.

The following examples will elaborate on the negative training definition.

- Any viewable aircraft instrument or display presentation that is not located in the same place, within a defined tolerance, as it is located in the real world.
- Any weapon simulation, sensor simulation, aircraft simulation or electronic warfare simulation that does not perform, within a defined tolerance, as it would in the real world.
- Any interface system that does not perform, within a defined tolerance, as it would in the real world.
- Any action required by participant individuals during an exercise that is inconsistent with real-world employment of tactical systems.

The simulations used during each training mission shall be validated to ensure that they will not cause any negative training.

Man-machine interfaces shall be user friendly with emphasis on simple, intuitive operation. System controls and displays shall emphasize logical operation by personnel using equipment with minimum operator training.

The flight simulator user interface shall be designed to minimize computer computer workout. Operator errors, such as incorrect commands or out-of-limits inputs, shall be detected by the software and shall not cause abnormal or premature cessation of a system operation.

The flight simulator shall be designed such that manpower requirements shall be minimized. A flight simulator plan shall determine the appropriate manpower and associated skill levels and training necessary for the program operation and maintenance.

The flight simulator logistics human factors shall include the following.

- Quality assurance design approach that shall ensure all products and services are easy to maintain.
- Reliability design of components shall ensure that the human operation of the total system is easily understood.
- Maintainability design of components shall allow rapid identification and repair of component failures.

The Systems Engineering human factors element in the design and development of a Research or Trainer Flight Simulator includes human factors experimental designs, conducted to analyze a specific potential problem with the product.

Human Factors experimental designs, conducted on the F-14 Aircraft Flight Simulator, provided mission type exercises that can be used to analyze a particular design or development issue. Some of the research experimental designs that were pursued include the following.

1. Evaluation of the F-14 Aircraft bomb display arrangement architecture, replacing the A-6 Aircraft arrangement.

2. Evaluation of the F-14 Aircraft spins display system.

3. Evaluation of a variety F-14 Aircraft head-up display architectures.

Human Factors experimental designs, conducted on the LAMPS Anti-Submarine Flight Simulator, provided mission type exercises that can be used to analyze a particular design or development issue. The LAMPS SH-2F / SH-60 Aircraft provide sensors that are used to search and detect an enemy submarine. Some of the research experimental designs that were pursued include the following.

1. Evaluation of aircraft avionic information on typical instrumentation versus a television monitor.

2. Evaluation of the best circle size of a television cursor, used to hook objects.

3. Evaluation of the best keyset legend methodology (Full word or Abbreviation).

4. Evaluation of submarine search exercises, using various buoy types (CASS, DIFAR, LOFAR) of avionics, used to detect location of submarine.

5. Evaluation of providing a tactical display to a pilot, in addition to the tactical crewperson.

6. Evaluation of providing various aircraft keyset architectures.

7. Evaluation of multi-sensor (radar and infra-red sensors) artificial intelligence design approach software.

8. Evaluation of an Instructor Operator Station (IOS) manual control (side arm controller versus space ball controller).

12.4.6 Flight Simulator Security

The Security Assurance Requirements Document (SARD) specifies functional and assurance security requirements for a Research or Trainer Flight Simulator. The SARD requirements have been derived from the ISO / IEC 15408 Document. The SARD requirement shall be specifically and formally traceable throughout the program.

A Research or Trainer Flight Simulator include a number of hardware components, models and interface systems that require security evaluation. The flight simulator tactical environment aircraft models, weapon models, radar models and electronic warfare models include numerical information that is classified. Flight Simulator cockpit displays present radar range data and launch acquisition range (LAR) data, which are classified. Flight Simulators include interface systems which may transfer selective classified information over the interface and therefore must provide security encrypted handling.

The Flight Simulator Program must segregate its classified information and assure that only authorized users are provided access to any classified information.

The Flight Simulator Program must establish security-relevant mechanisms that provide access rights, proper labeling and transmittal flow control.

The Flight Simulator Program must establish security-relevant accountability mechanism that provide for identification and audit of users of the classified material.

The Flight Simulator Program must be certified and accredited for multi-level security mode of operation, in accordance with the DoD Information Technology Certification and Accreditation Process (DITSCAP, DODI 5200.40). The DITSCAP requires verification and validation of a set of specified security requirements, which shall be defined in the requirements of the SARD.

To allow the public to tour the Research or Trainer Flight Simulator the classified information is varied with non-classified data.

To document the verification and validation classified testing that the public is allowed to view, the specific aircraft, weapon and radar are provided with a number instead of their specific name.

The Next Generation Threat System (NGTS) Program, which included the F-18 Trainer tactical environment, was generated in a classified facility and the System Requirements Document (SRD) and configuration management of the requirements; software and discrepancy reports were all classified.

12.4.7 Flight Simulator Logistics

12.4.7.1 Introduction

Flight Simulator logistics provides program supportability which includes program documentation, program material parts, spare parts, program maintenance, maintenance analysis, program testing, testing equipment, and configuration management. Presented below are details involved in each of these logistics efforts.

12.4.7.2 Logistic Documentation

The Capability Maturity Model Level 2 and 3 documentation is presented in Table 2.2.1 of the Software Engineering Life Cycle Process and repeated below for the design and development of a Research or Trainer Flight Simulator, with some additions.

- System Requirement Document (SRD)
- Software Development Plan (SDP)
- Configuration Management Plan (CMP)
- Software Quality Assurance Plan (SQAP)
- Project Training Plan (PTP)
- Software Design Document (SDD)
- Interface Design Document (IDD)
- Math Model Report (MMR)
- Requirements Traceability Matrix (RTM)
- Users Manual
- System Test Plan (STP)
- System Test Procedure (STP)
- Software Test Procedure (STP)
- Acceptance Test Report (ATR)
- Validation Test Report (VTR)

Other documents may be included depending on the specific program needs.

12.4.7.3 Program Materials, Spare Parts, Maintenance and Maintenance Analysis

The complete flight simulator program material list must be identified, as detailed in the Figure 12.4.1.1 Flight Simulator Hardware Block Diagram, including the item name, vendor part number, federal supply code of the manufacturer, quantity required, price, and possibly the mean time between failure (MTBF), and production lead time for each item.

In addition, a program material spare parts list needs to be generated with the same information that is detailed above. Spare parts need to be supplied for material that has a low mean time between failure values.

The material mean time between failure values can sometimes be determined through maintenance analysis. I found that keeping good laboratory documentation is the best way to determine material endurance. Documenting the laboratory down time, the reason for the downtime, the procedure to solve the downtime problem, and the time required. A trouble shooting guide should be generated from laboratory documentation.

A Mean Time Between Failure (MTBF), in hours, value needs to be identified for each major Flight Simulator component, when operated under the worst-case environmental conditions.

A Mean Time to Repair (MTTR), in hours, values needs to be identified for each major Flight Simulator component, when operated under the worst-case environment conditions.

A Mean Time Between Calibration (MTBC), in hours, values needs to be identified for each major Flight Simulator component, when operated under the worst-case environment conditions.

All subsystems shall conduct Built-In-Test (BIT) upon power-up, when commanded by an operator and in the background while the system is operational. Power-up and operator commanded BIT shall isolate failure, and not degrade the systems while under test. Self-diagnostics shall provide a mission ready go or no-go indication to the crew. Faults shall be logged to provide the operator the capability to display the BIT results. BIT shall have the capability to fault detect and isolate failures to a 90% level.

12.4.7.4 Program Testing

The program shall certify the technical accuracy and adequacy of the program products through verification and validation tests. Verification is testing the product against the design, while validation is testing the product against the real world performance values. Program testing will be completely described in the performance section of this book.

12.4.7.5 Program Configuration Management

Program Configuration Management is required for the program requirements, program software, and the program performance discrepancy reports. The Configuration Management operation is usually done with an industry software application. There are a number of good programs, including the following.

1. IBM Rational Clear Case (Software), Clear Quest (Performance) and Requisite Pro (Requirements).
2. Telelogic DOORS

As an example, the F-18 Trainer Next Generation Threat System (NGTS) tactical environment utilized the IBM Rational Clear Case, Clear Quest, and Requisite Pro.

The Next Generation Threat System (NGTS) tactical environment was designed and developed by the Navy Base, in Patuxent River, Maryland, the Air Force Base, in Mesa, Arizona and the Air Force Base, in Albuquerque, New Mexico. The development requires configuration management coordination between the three facilities. Presented below, in Figure 12.4.7.5.1, is the Configuration Management Network Topology.

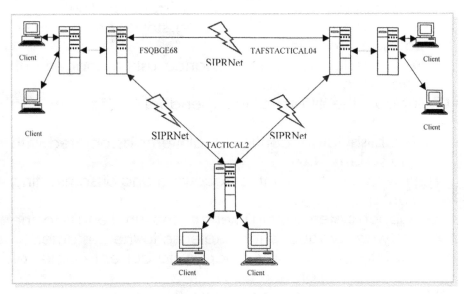

Figure 12.4.7.5.1 Configuration Management Network Topology

The Network Topology Intel Xeon 1.8 GHz Server includes a Windows NT 2000 Operating System, with a memory of 523 MB.

The Central NAS Patuxent River Configuration Management responsibilities, for the NGTS Program, included the following efforts.

1. Establish, maintain and enforce configuration management policies and procedures.
2. Train users to use the selected tool.
3. Maintain version control; for all work products.
4. Manage the change control process to track all reported problems, their resolution and implementation status for software products under configuration control.
5. Provide status accounting reports to the Configuration Control Board (CCB), and update the status accounting databases to reflect CCB decisions.
6. Provide a mechanism for allowing and controlling changes to products under configuration control.
7. Provide a mechanism for entering, storing and providing status information.
8. Perform Clear Case maintenance, user management, backups and recovery.
9. Support the efforts of Independent Verification and Validation (IV&V) audits.
10. Establish library control for all items associated with the system.
11. Provide central point for receiving and disseminating controlled work.
12. Support system testing to ensure the functionality of the software product prior to release to the customer.
13. Maintain and control access to current copies of documentation and code.
14. Distribute approved configuration.
15. Generate Configuration Management reports.

The Supplementary Mesa Air Force Base and Albuquerque Air Force Base Configuration Management responsibilities, for the NGTS Program, included the following efforts.

1. Contribute to the establishment, maintenance and enforcement of the agreed upon Configuration Management policies and procedures.

2. Train users to use the selected tool.
3. Monitor version control for work products.
4. Maintain current copies of documentation and code.
5. Provide current access control information to the Central Configuration Management office.
6. Prepare, control and disseminate copies generated from the configuration items distributed from the Central Configuration Management office.
7. Support the change control process: Report deficiencies and improvements to the Central Configuration Management office that is necessary to complete development directives and initiatives.
8. Perform Clear Case maintenance, user management, backups and recovery.
9. Contribute to and support the maintenance of the work products.
10. Support system testing as required ensuring functionality of the software product.
11. Provide inputs to the Configuration Management reports.

12.4.8 Flight Simulator Safety

A System Safety Program, in conjunction with the Software Safety Program, shall be pursued to ensure that the program hardware, software, weapons, interfaces, and human factors, does not adversely affect the Trainer Flight Simulator performance, and the support equipment, avionics and all other ancillary equipment associated with the program system. In addition, the safety of the flight simulator crewpersons is of particular concern.

1. Platform modifications for accommodating installation of the program equipment shall not degrade the safety of the equipment.

2. Safety hazard analysis shall be executed, in accordance with MIL-STD-882D

3. Protection against electrical shock hazards, use of toxic materials, mechanical equipment design and other personnel safety considerations shall be in accordance with MIL-STD-454.

4. Program equipment intended for shore facilities shall meet the electrical installation requirements of NFPA-70 (National Electrical Code).

Each one of the Systems Engineering elements, including the hardware, software, modeling, interface, human factors, security and logistics, should be analyzed to determine if there are any safety problems with their design.

12.5 Flight Simulator Product Stories

During my 48 years of Department of Defense service I have worked on the design and development of the following Research and Trainer Flight Simulators. The F/A-18 Aircraft Trainer Flight Simulator, the F-14 B/A Aircraft Trainer, the F-14D Aircraft Trainer Flight Simulator, the F-14D Research Flight Simulator, the SH-2F Helicopter Trainer Flight Simulator, the F-14A Aircraft Dynamic Flight Simulator, the Crewstation Design Research Flight Simulator, the SH-2F / SH-60 Helicopter LAMPS Flight Simulator, the F-111 Aircraft Research Flight Simulator, and the CH-53 Helicopter Research Flight Simulator.

Presented below are a couple of stories regarding some of these flight simulator products.

1. F-18 Aircraft and F-14B/A & D Trainer Flight Simulator Product (1986-2006)

The F-18 and F-14B/A & D Trainer Flight Simulator contract did not provide the specific training requirements and therefore the final product had to be used with the specific hardware, software, modeling, interface, human factors, security, logistics and safety provided. It was determined in the product testing that there were a great deal of limitations to the trainer flight simulator product provided.

The F-18 Trainer and F-14B/A & D Trainer utilization were continually thought about after the flight simulator products were provided. The Trainer technical limitations were analyzed after the products were provided and not more carefully analyzed before the product design and development began.

The fidelity of the F-18 Trainer Next Generation Threat System (NGTS) aircraft models is one example of the product that has to be radically changed to be useful for any close-in combat training exercises. The initial NGTS Tactical Environment aircraft model did not include variable turn rates, climb rates and acceleration rates; therefore the proper aircraft tactics could not be generated

Another example was the limitation of the jammer model types for the F-14B/A & D Trainers that did not allow Electronic Warfare (EW) training with the latest jammer types. This problem was also a limitation of the F-18 Trainer.

Coordination of the past F-18 Trainer tactical environment with the new F-18 Trainer NGTS tactical environment is a major limitation.

2. Tacair Systems Development Facility (TSDF) Product (1988-1990)

The Tacair Systems Development Facility (TSDF) provided an F-14D Aircraft Trainer that analyzed the installation of bombs being incorporated into the F-14 Aircraft and a new Head-Up Display (HUD) system that was being considered for the F-14 Aircraft. The development of the facility was being conducted by four groups. The crewstation and the software was designed and developed by the Veridian Company. The indiscrete and outdiscrete signal design was designed by an in-house engineer from our division. The cockpit computer J-Box system was being designed by the Vitro Company. The D/A and A/D Interface was being designed by the Computer Department, through their contractor DSG Company.

The TACAIR Department Captain stopped by my desk just about every day, in early 1990, to ask me when the TSDF Facility would be finished. Four of the five components were complete and the only component that still needed completion was the D/A & A/D Interface. As Program Manager, I tried in every way I could to get the Computer Department to work on their effort, but had difficulty in getting them to finish the effort. The Computer Department went to upper management and said it was my fault that they were not finishing their effort, and they wanted to take over the program. This situation is a difficult situation for a Program Manager. The TSDF Facility was finally completed in the next fiscal year, when the D/A & A/D Interfaces were completed. The final F-14D Aircraft Research Flight Simulator turned out to be a valuable product.

One aside, is that I personally ran out of program funding for the last three months of the fiscal year 1990, and had to beg for funding from a friend, Al Gramp, to allow me to stay on the program for the year. I generated two reports, related to some of Al Gramp's TACTS Range Program. At the end of the year, I checked the statement of the funding spent by the Computer Department for the D/A & A/D interfaces, and found that they never spent any of the funding that I had given to them; I guess because they had enough funding for their yearly efforts and didn't really need my funding. The F-14D Aircraft Research Flight Simulator, located in the TACAIR Systems Development Facility (TSDF), was taken over by the TACAIR Department, shortly after the facility was completed. During the development of the TSDF, I traded the Analog Computer Facility to obtain funding to construct the TACAIR System Development Facility. I designed and developed the TACAIR System Development Facility, including the System Engineering design of the F-14D Aircraft Flight

Simulator. I generated dozens of contracts for all of the F-14D Aircraft Flight Simulator equipment, including the "out-the-window" visual display system, flight control system, displays, instruments, panels, switches, and indicators. I managed the design and development of the entire F-14D Aircraft Flight Simulator. The TACAIR Department liked the final product so much that they took it over and my management did not raise any objections.

Takeovers of engineering facilities at the Naval Air Development Facility were a common occurrence over the years. Many other engineering friends lost their facilities in an organization takeover.

During the design and development of the TACAIR Systems Development Facility I was given $75,000 to provide a throttle control loader system. I generated a contract Request for Proposal (RFP) and went out to industry to design and develop the system. Several companies responded, but none of the proposals were acceptable. The contract Request for Proposal (RFP) was sent out into industry a second time, and again none of the proposals were acceptable. In desperation, as the Program Manager, I designed the throttle control loader system, at night at home, and built the system in our Naval Air Development Center machine shop. The system worked out very well and I received a patent for the design. The major features of the design included the following,

- A throttle stick that could be moved forward through the normal aircraft speed positions and then moved to the left & forward for the afterburner position.

- A voltage indication between -10 volts to +10 volts for all speed positions.
- A friction device to allow the throttle sticks to be located in any speed position.

3. SH-2F Helicopter Trainer Flight Simulator Product (1985-1987)

The SH-2F Helicopter Update Program included the replacement of the simulator computer, re-hosting of the simulator software, replacement of the simulator flight control system and the generation of a new engine aerodynamic software and flight control system software. The program team was very capable of providing an excellent product, and after the flight simulator updates were completed and checked out the final product was very well done.

The problem with this program was the short schedule identified to complete the program. The Captain wanted the program to be completed faster than the program team could finish the program. During installation and checkout, my engineers were required to work a seven day week. As Program Manager, I was called aside one day and asked to explain why one of my engineers had taken a one hour lunch and why this engineer had taken off on a Sunday afternoon. I explained that my engineers were working very hard and they could not be expected to work every minute of a week. The Captain listened to my defense of my workers and from that point of time accepted my management of the program.

The tradeoff of the program cost, schedule, and quality of the product must be discussed during the design and development of a program.

4. F-14A Aircraft Dynamic Flight Simulator (1982-1985)

The F-14A Aircraft Dynamic Flight Simulator provided an F-14A Aircraft Research Flight Simulator that could be flown with 15 G's of force in the X, Y and Z directions, to study the affects of an aircraft spin. All the components of the dynamic flight simulator had to be designed to accommodate 15 G's of force. The Dynamic Flight Simulator Program was very successful, and several spin excises were conducted.

In 1985 Vice President George Bush visited the Naval Air Development Center and toured the Dynamic Flight Simulator. The Secret Service visited the week before the Bush visit and we were told the he could not travel upstairs on the elevator, which we had just painted. We had the stair well painted, but you could still smell the fresh paint when he arrived. The Vice President flew our flight simulator, just outside the centrifuge area, and performed very well. At one point in time he said that he had to go to the Men's Room, which we did not think of having painted or even cleaned. I was proud to have such a distinguished visitor see one of my engineering programs.

Demonstration of a program to distinguished visitors quite often provides additional funding for the program.

5. Crewsystem Design Flight Simulator (1975-1996)

The Crewsystem Design Flight Simulator was a generic flight simulator used to introduce new technology to an aircraft. Over the years the liquid crystal display was introduced, the all crewstation displays was introduced, which later was the F-18 Aircraft crewstation concept and the interface with other remote flight simulators were introduced, which is now used in a major battle scenario exercise.

My Terrain Model Gantry System was located in the Crewsystem Design Flight Simulator Facility. The Terrain Model Gantry System, illustrated in Figure 12.5.1.6.1 and Figure 12.5.1.6.2, is used to provide an "out-the-window" visual display. This type system was used before the design and development of a computer generated image visual display system. I designed and developed the Model Gantry System in the 1973-1975 time periods, and when the new Software Branch Head took over our organization in 1975, he had no interest in the Model Gantry System. As a matter of fact, I had just traded for a high bay space, with a 20 foot high ceiling, to locate the Model Gantry System, and he decided to use this space as an office space for some of his software engineers.

Providing interest in your program product to your management is an important part of providing a successful product

6. SH-2F / SH-60 Helicopter LAMP Flight Simulator (1970-1990)

The SH-2F / SH-60 Helicopter LAMPS Flight Simulator provided the first device for the anti-submarine warfare system on a helicopter aircraft. The LAMPS Flight Simulator provided a facility to pursue many experimental design exercises to evaluate new anti-submarine technology. One of the experimental designs was to evaluate the Cass Buoy capability of finding a submarine. The Cass Buoy is a passive system that detects the range between the buoy and the submarine. Using two Cass Buoys the submarine track can be determined, but the submarine direction is not determined.

The experimental design provides for the LAMPS Helicopter to fly a path around the gaming area until the CASS Buoy detections are obtained and a submarine path is evaluated. A Canadian LAMPS Helicopter pilot used his intuition of the submarine path and just after the experimental design was initiated he went directly for the submarine and found it immediately every time. Of course, the purpose of the experimental design was not satisfied, but he proved to be an excellent anti-submarine pilot.

Detailing the specific experimental design objectives need to be presented to all of the persons involved in the exercise.

7. F-111 Aircraft Research Flight Simulator (1967)

The F-111 Research Flight Simulator was being designed and developed when the program ran out of funding.

When you are out of funding the program ends, so assurance that sufficient funding to complete a successful product is very important, but never completely possible.

8. CH-53 Helicopter Flight Simulator (1965-1968)

The CH-53 Helicopter Research Flight Simulator was being designed and developed to introduce new technology into the aircraft. An aircraft cockpit display that introduced a real world vertical image to the pilot was analyzed.

A cockpit Map Display was also analyzed. I designed and developed the system, using a transport device to provide paper maps with a number of different scales to the crewstation person. A hardware aircraft bug was designed to fly over the paper map to indicate to the crewstation person the exact location of his real world position. The Map Display actually worked and at one point in one of the experimental design exercises an engineer asked if the aircraft bug was at the correct longitude/ latitude position and realized that the aircraft bug was directly over the correct map display position. I have a patent on the Map Display.

Today, a high fidelity television image is presented to a crewstation person with a map image and an aircraft bug image, but of course my design was provided thirty five years ago.

Chapter 13 Flight Simulator Process

13.1 Introduction

Systems Engineering transforms an operational need into a final product. The Systems Engineering Life Cycle Process has changed throughout the years. A generic Program Life Cycle Process is presented below, in Figure 13.1.1

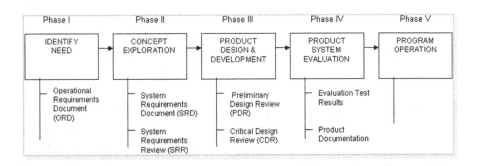

Figure 13.1.1 Program Life Cycle Process

13.2 Program Process

The F-18 Trainer Flight Simulator and F-14B/A & D Trainer Flight Simulators were designed and developed through government contracts to contractors, so the program process pursues this approach.

The F-18C Trainer and F-14B/A & D Trainer Programs were initiated from the NAVAIR through an Operational

Requirements Document (ORD). The Naval Air Warfare Center Training Division, in Orlando, Florida then, generated a System Requirements Document (SRD), from the ORD, with the technical support from several other Navy bases, including The Naval Air Warfare Centers in Warminster, Pennsylvania, Patuxent River, Maryland, PT MUGU, California, and China Lake, California. The System Requirements Document was put into a Contract Request for Proposal (RFP) Statement of Work and various contractors bid on building the F-18C Trainer and F14B/A & D Trainers. Each of the contracts awarded were a fixed price, low bidder contract.

The new F-18C Trainer Flight Simulator was designed and developed, in the mid 2000 timeframe, by L3 Communication Company, Dallas, Texas. L3 Communication Company is basically the Singer Simulation Company, formerly from Binghamton, New York and the Hughes Aircraft Company, formerly from Los Angeles, California, who developed many of the former F-18 and F-14 Trainer Flight Simulators. The F-18C Trainer Flight Simulators are located in the Naval Air Station (NAS) Oceana, Virginia and the Naval Air Station (NAS) LaMoore, California.

The F-14 D Trainer Flight Simulator was designed and developed, in the early 1990's timeframe, by McDonnell Douglas Corporation, McDonnell Douglas, in St Louis, Missouri and AAI Corporation, in Cockeysville, Maryland.

The updated F-14 B/A Trainer Flight Simulator was designed and developed, in the mid 1990's timeframe, by Grumman Aerospace Corporation, in Great River, New York.

The F-14 B/A & D Trainer Flight Simulators were initially located in the Naval Air Station (NAS) Miramar, California and the Naval Air Station (NAS) Oceana, Virginia. After the Base Re-Alignment and Closure (BRAC) Committee closed the NAS Miramar trainer facility in 1996 and moved all the trainers to NAS Oceana.

The F-18C Trainer and F-14B/A & D Trainer Program Process followed the Program Life Cycle Process, detailed in Figure 13.1.1. This process included the System Requirements Review (SRR), the Preliminary Design Review (PDR), the Critical Design Review (CDR) and after the contractor designed and developed the trainer flight simulator product, the government pursued the operational and performance evaluation of the trainers.

Each contract, after a period of time, identified the software process of utilizing the Capability Maturity Model (CMM) Level 2 and 3 for the majority of the software development processes. The details of this process is identified in Chapter 4 Table 4.2.1.1 Capability Maturity Model (CMM) Level 2 & 3 Process Questions.

All of the Naval Air Development Center Research Flight Simulators were basically developed using an in-house design and development process. These Research Flight Simulators include the CH-53 Helicopter Research Flight Simulator, the F-111 Aircraft Research Flight Simulator, the SH-2F /SH-60 Helicopter LAMPS Research Flight Simulator, the F-14A Aircraft Dynamic Flight Simulator, the F-14D Aircraft Research Flight Simulator, the Crew Systems Research Flight System, and the F-14A Aircraft Air Combat Maneuvering (ACM) Research Flight Simulator. In addition, an Image Generator gantry system was designed and developed to support the flight simulator visual display systems. The Research Flight Simulators were designed and developed using in-house engineers and local contractor engineers (about 50% of each), in-house machine shops, electrical shops and installation shops and contractor shops (about 50% of each).

This book will concentrate on the government contracting process for the design and development of a Trainer Flight Simulator, such as the F-18 Trainer and the F-14B/A & D Trainer Flight Simulators.

13.3 Flight Simulator Process Stories

During my 48 years of Department of Defense service I have worked on the design and development of the following Research and Trainer Flight Simulators. The F/A-18 Aircraft Trainer Flight Simulator, the F-14 B/A Aircraft Trainer, the F-14D Aircraft Trainer Flight Simulator, the F-14D Research Flight Simulator, the SH-2F Helicopter Trainer Flight Simulator, the F-14A Aircraft Dynamic Flight Simulator, the Crewstation Design Research Flight Simulator, the SH-2F / SH-60 Helicopter LAMPS Flight Simulator, the F-111 Aircraft Research Flight Simulator, and the CH-53 Helicopter Research Flight Simulator.

Each of the Naval Air Development Center Research Simulators design and development was accomplished in-house, so no Preliminary Design Reviews (PDR) or Critical Design Reviews (CDR) occurred to evaluate the product as the process pursued. Functional Management Leaders provided some design reviews which were somewhat similar to PDR's and CDR's, but not nearly as detailed as the actual PRD or CDR.

I attended every F-14B/A and D and F-18C Trainer Flight Simulator System Requirements Review, Preliminary Design Review and Critical Design Review. In humor, I often said that you needed to bring five lawyers to every design review, because during each design review something occurred that strongly affected the design and development of the flight simulation product. The trainer contracts were all low bidder, fixed priced, contracts, so as soon as the contract was awarded the contractor was looking for how to obtain more funding on the contract. Presented below are a couple of stories that occurred during these design reviews.

Enclosed are a couple of stories detailing the process of the Trainer Flight Simulators.

1. F-18 Trainer Tactical Critical Design Review (2004)

The F-18 Trainer Critical Design Review (CDR) described using the new Next Generation Threat System (NGTS) tactical environment with the old F-18 Trainer tactical environment, but when questioned they admitted that they had no coordination software. The one tactical environment aircraft would not know the location of the any aircraft in the other tactical environment. The two tactical environments had to be used separately.

2. F-14B/A Trainer Critical Design Review (1995)

On the F-14B/A Trainer contract, the government was tasked to provide a government furnished special cockpit panel, by the Critical Design Review date. The sub-contractor building the special cockpit panel was behind schedule at the Critical Design Review. The F-14B/A Trainer contractor used this issue to identify that this late panel delivery caused them to be behind schedule and they needed a great deal of funding to complete the trainer contract. The government offered $2 million and the contractor asked for $4 million, to complete the program. Later in the negotiations, the government offered $4 million and the contractor asked for $8 million. I don't know the final resolution to the contract negotiations, but the incident shows the difficulty of providing a successful product.

3. F-14B/A Trainer Preliminary Design Review (1994-1995)

At the F-14B/A Trainer Preliminary Design Review (PDR), I asked if the contractor had exercised the tactical environment aircraft models on the selected computer, remembering that on the F-14D Trainer contract the required number of high fidelity aircraft models were not provided. The contractor said that they would exercise the aircraft models before the Critical design Review. At the F-14B/A Trainer Critical Design Review (CDR), the contractor said that they were identifying a new computer system, because of their aircraft model exercise.

4. F-14D Trainer Preliminary Design Review (1988)

During the F-14D Aircraft Trainer Flight Simulator Preliminary Design Review (PDR) it was identified that the specific aircraft, weapons and radar models required were not specifically identified, in the F-14D Trainer contract statement of work. A general list was provided in the contract package. The government team met at night after the first day of the PDR to specifically determine the specific tactical environment model list. The list was generated by the military persons attending the meeting, but I was able to listen to their discussions and realized how important having the correct models provided in the trainer flight simulator would produce a good product. I was required to provide the TACTS Range weapons models for the trainer and the contractor date for this provision was the PDR. Weapon models that were identified that night that had not already provided to the contractor were not included in the current F-14D Trainer contract. The contractor required more funding to add these models to the F-14D Trainer tactical environment.

5. F-14D Trainer Preliminary Design Review (1988)

The F-14D Trainer required 16 high fidelity aircraft models. The contractor found that the computers selected could not provide more than 4 high fidelity aircraft models. The government allowed this change, which I did not agree with. The low fidelity models were then presented to be included in the air combat maneuvering (dog fight) exercises. The low fidelity aircraft models were designed with only 3 degrees of freedom. The limitation of not having a roll, pitch and yaw capability "within visual range" was unacceptable and the PDR determined that the low fidelity aircraft models needed to be redesigned.

6. F-14D Trainer Preliminary Design Review (1988)

The F-14D Trainer required 25 missiles to be in the air at one time. The contractor found the computers selected could not provide more than 9 missiles in the air at one time. The government allowed this change, which I did not agree with. I also thought that you could trade off this change for requiring the contractor to allow all of the TACTS Range models identified at night during the PDR, but this was not done. The contractor obviously had a stronger will than the government.

Chapter 14 Flight Simulator Performance

14.1 Introduction

Presented below, in Figure 14.1.1 is a typical program design and development process, which presents the tasks needed to evaluate the performance of the program product.

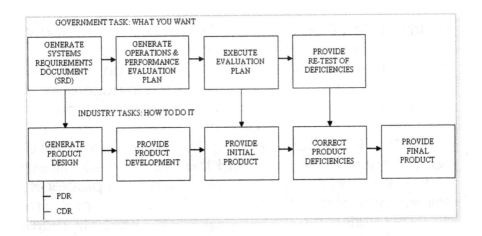

Figure 14.1.1 Program Performance Evaluation Process

14.2 Flight Simulator System Performance Evaluation

The F-18 Trainer Flight Simulator and F-14B/A & D Trainer Flight Simulators were evaluated in accordance with the process illustrated in Figure 14.1.1. The government generated a Trainer Test Procedure and Results Report (TTPRR) to evaluate the trainer operational capability and a Validation Test Procedure and Results Report (VTPRR) to evaluate the trainer performance capability.

In addition, the utilization of military aircraft crewpersons exercising the trainers capabilities to evaluate the operational and performance of the trainer is significant.

The Trainer Test Procedure and Results Report are about 12 volumes long, with hundreds of pages in each volume. Each element of the trainer is exercised with an operation and determination if the operation works in accordance with the test results in the report. For example, a button is pushed and the resultant affect is observed. The Trainer Test Procedure and Results Report usually take many months to execute and the problems identified take contractors many months to correct the problems. Deficiency Reports (DR) are generated for each problem identified during the trainer evaluation. Typically, more than a thousand Deficiency Reports are generated during the trainer evaluation.

Presented in Table 14.2.1, are a couple of pages of the F-14 B/A Trainer Test Procedure and Results Report (TTPRR) specific step-by-step procedures. The Table 14.2.1 Trainer Mission Setup is an example of what is provided in a Trainer Procedure and Results Report.

Table 14.2.1 Trainer Mission Setup

Procedure Step	Expected Result
1. At the IOS keyboard, press the Change Modes key.	The change modes page is displayed on the IOS touch screen.
2. On the IOS touch screen, touch the Plan legend.	The prompt for plan mode access code is Displayed on the IOS touch screen.
3. Enter the Plan Mode Access code.	The Plan Mode menu appears on the IOS touchscreen.
4. On the IOS touch screen, touch the Mission Setup legend.	The Mission Menu appears on the IOS touch screen.
5. Add the number of missions which are Currently defined on the device by adding The numbers which appear next to the section titles.	If the total number of missions is less than 40, the next two steps need not be performed. If all 40 missions are defined on the device, perform the next two steps.
6. On the IOS touch screen, select a box next to a category displayed touch delete selected mission, then touch yes.	One mission is deleted, thus allowing for the addition of another mission.
7. Touch the Mission Menu legend on the IOS touch screen.	The Mission Menu page is displayed on the IOS touch screen.

8. On the IOS touch screen, touch The box to the left of the Test legend. — The Mission Index Menu for Test Missions Appears on the IOS touch screen.

9. On the IOS touch screen, select 4. — Mission 4, which begins with the aircraft In flight, is selected.

10. Touch the Add New Mission legend. — The Mission Menu is displayed on the left IOS Touch screen and the Mission Summary is displayed on the right IOS touch screen. The prompt for the Mission Number is displayed on the IOS keyboard.

11. At the numeric keypad enter the Number listed for the upper limit. — Test Mission Index is displayed on the IOS touch screen.

12. Touch the Mission Menu legend on The IOS touch screen — The Mission Menu page is displayed on the IOS touch screen.

13. On the Mission Menu touch the View / Modify Selected Mission Legend. — The following View / Modify category selections are displayed on the IOS keyboard: Radio, Surf Fac, Cont Align, Air Field, Envir, Stores, CV, Map, ACFT Data, Target Threat, MALF.

14. At the IOS keyboard, select: Envir Weath — The IOS touch screen displays the weather page.

15. At the touch screen, touch S / L Temp And enter 42 — The new Sea Level Temperature is modified and updates the IOS touch screen correctly for this new Mission Setup.

Pilot-Only Independent Mode

16. On the IOS keyboard, key Change Modes.

The Change Modes page is displayed on the IOS touch screen.

17. On the IOS touch screen, touch The Training legend.

The Training Mode setup page is displayed On the IOS touch screen.

18. On the IOS touch screen, touch The Pilot-Only Independent legend.

The corresponding selection is indicated.

19. On the IOS touch screen, touch the Initialize to Selected Mission legend.

The Control Alignment Page is displayed on The IOS touch screen. The Reset Key on the IOS keyboard is flashing.

20. On the IOS touch screen, touch the Initialize to Selected Mission legend.

The Control Alignment Page is displayed on the IOS Touch screen. The Reset key on the IOS keyboard is Flashing.

21. Reposition switches in the cockpit listed on the Control Alignment page To correct positions, then select Continue Mission Initialization legend.

When a switch is set to the required position, it is removed from the display. When the last switch is set to the required position, the Reset key extinguishes on the IOS keyboard and the freeze Key begins to flash.

22. At the IOS keyboard, select: Envir Weath.

The IOS touch screen displays the weather page.

23. At the IOS touch screen, look at the box next to S/ L Temp.

The Sea Level Temperature of 42 which was modified in the Mission Setup is displayed on the IOS touch screen correctly for this new Mission Setup.

24. On the IOS keyboard, press the Acft Data key, then the Acft Param key. Reset selections available.

The Aircraft Data menu is displayed with Restore Ins Align System Reset, and Emer Gear Handle

25. On the IOS keyboard, press the Run / Freeze key.

The Run legend on the Run / Freeze key is highlighted.

26. Pilot mains current altitude and Airspeed or engages auto pilot.

Pilot flight conditions are normal.

27. Verify that the System Reset selections are functional by touching the corresponding selection on the IOS touch screen.

The Mission Computer are reset. The default displays are shown on the VDI and HUD as follows: HUD: Basic Takeoff VDI: Basic Takeoff

28. Touch Run / Freeze on the IOS keyboard.

The Freeze legend on the Run / Freeze key is highlighted.

29. At the IOS keyboard, press the reset key.

A Yes / No prompt is displayed.

30. Touch Yes at the IOS touch screen.

The Mission Menu page is displayed on the IOS touch screen.

14.3 Flight Simulator Model Evaluation

The Trainer Flight Simulator model Independent Verification and Validation (IV&V) evaluation includes the evaluation of the entire tactical environment, including the aircraft models, weapon models, radar models, sensor models, electronic warfare models. Initially, these models are evaluated individually, usually in an off-line computer. They are evaluated first operationally to see if they simply function as you would expect them to operate in the real world. Then the model performance is evaluated to see if the model performs specifically as the real world system. The operational tests measure the specification words and the performance tests measure the specification numbers. This book will provide, in great detail, the individual performance evaluation of the aircraft, weapons, and radar models.

Verification evaluates if the model matches the program design and validation evaluates if the model matches the real world performance.

The tactical environment models then need to be incorporated into the Trainer Flight Simulator and be evaluated as a system, using the aircraft missions to facilitate the evaluation. The generic missions include the following.

- Air-To-Aircraft Intercept
- Air-To-Ground Weapon Delivery
- Air-To-Ship Weapon Delivery
- Integrated Air Defense System (IADS / SAM Sites)
- Electronic Warfare
- Outer-Air-Battle (OAB)
- Air Combat Maneuvering (ACM)

In addition, the Trainer Flight Simulator evaluation needs to provide evaluation of the following tactical environment elements.

- Aircraft Tactics Rules
- Visual Priority Rules
- Skill Level Rules
- Groups
- Jammers
- Interactive Aircraft, Ground and Ships
- Aircraft Carrier Operation

All of the following tactical environment evaluation information will be for the F-18 Trainer Next Generation Threat System (NGTS) Tactical Environment.

This book will also provide details on the evaluation of the flight simulator aircraft engine performance.

14.3.1 Model Operational Evaluation

14.3.1.1 Model Operational Specifications

In general, the fight simulator model operational evaluation attempts to determine that the models operate in a realistic functional manner. The aircraft models need to fly in the correct direction, at the correct altitude, turn when they are directed to turn, fire a weapon when they are directed to fire, and in general perform as they perform in the real world.

The weapon models also need to fly in the right direction, at the correct altitude and in general perform as they perform in the real world.

The radar model needs to detect the target, acquire the target, track the target and provide a signal to fire a weapon at the target

The evaluation is subjective without specific performance numbers to specifically measure against.

14.3.1.2 Model Operational Evaluation

The following Next Generation Threat System (NGTS) tactical environment operational evaluation example is provided from the NASMP Block 1 Tactical Environment System Test Procedure Report, dated 3 November 2003.

The Test No/Title 1.1.1.1 F/A-18C Operational Evaluation presented below is an example of a typical trainer operational evaluation test.

Test No/ Title 1.1.1.1 F/A-18C

Test Objective: The objective of the test is to verify the maneuver, sensor and weapons capability of the F/A-18C entity type.

Initial Conditions: This test requires use of the f18ctest scenario, as defined in Appendix D-1.1.1.1

> NGTS Model is running on its appropriate computer.
> NGTS IOS is running on its appropriate.
> HLA Listener is running on its appropriate computer.

Step No.	Control Action	Expected Results
1	At the NGTS IOS, select and load The f18ctest scenario.	Loading appears on the IOS display for up to 2 minutes, and then the message "Freeze ox19, Scenario OK, OCC OC" appears in the MSG box.
2	At the NGTS IOS, select the map of U.S. button and select "Davis Monthan AFB" and then "TPC 1:500K"	Screen Zooms in on players
3	Select the Display Details control icon and enable the display of Routes.	The route points are displayed.
4	Select the Display Details control icon and enable the display of Beams.	The radar sensor beam and its sweep are displayed for the F/A-18C blue friendly aircraft.

5	Place the IOS cursor over the F/A-18C platform and observe the initial speed, course and altitude readouts.	The F/A-18C initial speed is indicated as 0 knots, the initial course is 140 degrees and the initial altitude is 3000 msl.
6	At the NGTS IOS select run by clicking the stoplight icon. The scenario begins and the platforms been moving.	The stoplight highlights green. The message "Running 0x11, scenario OK, OCC OK" appears
7	Place the IOS cursor over the F/A-18C platform and observe the speed and course readouts.	The F/A-18C speed is 400 knots and while the platform is moving from the initial route point to the second route point the course is 158 degrees.
8	Using IOS cursor, right click on the F/A-18C platform for sensor display.	Aircraft console window is displayed showing the sensor and mode status of the F/A-18C.
9	F/A-18C begins radar search	In the Aircraft Console window, the text showing the radar mode used in SCH box is highlighted green.
10	Once the F/A-18C platform has Passed the first route point and turned to the second route point, observe the Aircraft Console window.	In Aircraft Console window, the text showing the radar mode used switches from SRH to ACQ and then to TWS. The current mode is highlighted green.

11	Observe the speed and course readouts over the F/A-18C.	The F.A-18C speed increases to 504 knots and the course shifts to 285 degrees as the F/A-18C pursues the threat aircraft.
12	Observe the F/A-18C platform and watch for missile fire.	The F/A-18C shoots an AIM-7 missile at the target. In the Aircraft Console window, the text in the Fire Box is highlighted green while firing and missile is displayed on screen.
13	After missile intersects the target, place cursor over skull & crossbones.	Read out shows that target is dead.
14	Observe the Aircraft Console window.	After the missile has destroyed the target the Aircraft Console window displays that the aircraft switches back to TWS mode and the F/A-18C switches to FLYROUTE tactic.
15	Once the F/A-18C has completed it's turn back to the route points, select freeze by clicking the stoplight icon.	The stoplight highlights Red.
16	Select the Physical_Entity_State output of the HLA Listener and select the F18-01 player logged data for analysis.	The ploys of Lat/Long vs. Time and Course vs Time accurately reflect the scenario.

14.3.2 Model Performance Evaluation

14.3.2.1 Model Performance Specifications

The Trainer Tactical Environment Performance Specifications are presented in Table 14.3.2.1.1 The Performance Specifications is the information used to provide the Independent Verification and Validation of the trainer tactical environment. The performance specifications were generated over a ten year period from many of the world's tactical environment modeling experts. The Universal Threat System For Simulators (UTSS) Program, pursued from 1991 to 2001, was designed to provide a library of threat models and threat data.

This program provided the opportunity to discuss the evaluation of a tactical environment with the experts throughout the world. The Trainer Tactical Environment Performance Specifications were generated from these conversations and review of the information generated was also pursued with these experts.

Table 14.3.2.1.1 Trainer Tactical Environment Performance Specification

AIRCRAFT SIMULATION PERFORMANCE

AircraftPerformance Characteristics	Aircraft Simulation Fidelity (% Deviation from Real World Values)		
	High	Medium	Low
(1) Turn Rate Vs Altitude	+ / - 5%	+ / - 10%	+ / - 25%
(2) Energy Management (Turn Rate Vs Altitude)	+ / - 5%	+ / - 10%	+ / - 25%
(3) Climb Rate Vs Altitude	+ / - 5%	+ / - 10%	+ / - 25%

(4) Flight Envelope (Max / Min Speed Vs Altitude)	+ / - 5%	+ / - 10%	+ / - 25%
(5) Acceleration	+ / - 5%	+ / - 10%	+ / - 25%
(6) Database Values	Verification of values from reliable source		

WEAPON SIMULATION PERFORMANCE

Missile Performance Characteristics	Missile Fidelity (% Deviation from Real World Values)		
	High	Medium	Low
(1) Intercept Velocity	+ / - 5%	+ / - 10%	+ / - 25%
(2) Time of Flight	+ / - 5%	+ / - 10%	+ / - 25%
(3) Max / Min Range	+ / - 5%	+ / - 10%	+ / - 25%
(4) Failure Reaso	Equivalence	Equivalence	Equivalence
(5) Database Values	Verification of values from reliable source		

RADAR SIMULATION PERFORMANCE

Radar Performance Characteristics	Radar Simulation Fidelity (% Deviation from Real World Values)		
Nose to Nose & Nose to Tail	High	Medium	Low
(1) Detection Range	+ / - 5%	+ / - 10%	+ / - 25%
(2) Acquisition Range	+ / - 5%	+ / - 10%	+ / - 25%
(3) Track Range	+ / - 5%	+ / - 10%	+ / - 25%
(4) Fire Range	+ / - 5%	+ / - 10%	+ / - 25%
(5) Database Values	Verification of values from reliable source		

The Validation Test Procedure and Results Report is generated to exercise the specific performance of important elements of the trainer, such as the tactical environment, including the aircraft, weapons, radar, electronic warfare and the evaluation of the trainer aerodynamic and propulsion performance. For example, for an aircraft product, the real world aircraft maximum altitude, velocity versus altitude, acceleration versus altitude, turn rates versus altitude, energy management values versus altitude, and rate of climb values are evaluated.

Validation measures the real world parameter values against the product parameter values. Verification measures the design parameter values against the product parameter values. Verification compares the tactical environment parameter numbers against the real world parameter numbers, such as aircraft minimum and maximum speeds versus altitude. Verification analyzes the aircraft tactics, visual display priorities and skill level design.

The value difference tolerance determines if the product is high fidelity, medium fidelity or low fidelity. The Validation Test Procedures and Results Report provide a very valuable document that provides the overall product performance. The Validation Test Procedure and Results Report usually take several weeks to execute and the problems identified take contractors many months to correct. Discrepancy Reports (DR) are generated for each problem identified during the trainer evaluation. Typically, hundreds of Discrepancy Reports are generated during the trainer evaluation. A typical Discrepancy Report Form is presented in Figure 14.3.2.1.1

(BLACK INK ONLY)

DISCREPANCY REPORT NO: _____ RESPONSIBLE ENGINEER: _____

DEVICE:	S/W BASELINE NO:	DATE:
TEST: _____ BOOK: _____ PG: _____		TTPRR REV: _____
TEST NAME: _____		CAT/CODE: _____
ORIGINATOR: _____		TEC REPORT: _____

DEFICIENCY:

CORRECTIVE ACTION:

RECHECK:	PASS	FAIL	DATE	NAME
#1				
#2				

COMMENTS:

DEFICIENCY CLOSED:

NAWC-TSD: _____ DATE: _____

(NAWC-TSD SIGNATURE MUST BE PRESENT FOR CLOSURE)

Figure 14.3.2.1.1 Discrepancy Report Form

14.3.2.2 Generation of NGTS Reference Values

Initially, the NGTS Operational Requirements are exercised to determine if the aircraft, weapons, radar and electronic warfare models operate correctly. These tests show that the aircrafts and weapons fly in the direction specified, turn, climb, accelerate, and in general operate as the models should operate. These tests also show that the radar and electronic warfare models operate as the models should operate.

The model performance tests show that the aircraft, weapons, radar and electronic warfare models perform as the system performs in the real world. The real world values must be obtained and then the model is exercised to determine if the model performance is within the specification tolerance of the real world values. The following illustrations will present generic aircraft real world performance information.

Figure 14.3.2.2.1 presents the front page of the National Air Intelligence Center Engineering Report Analysis Document that the various figures of information are taken. Most of these reports are classified, but this specific report is unclassified, because the information is for a foreign version of an American aircraft. The data presented is somewhat devated from the actual real world specific data. In addition, all information regarding the specific aircraft is removed from each of the following figures.

Figure 14.3.2.2.2 presents an aircraft Altitude versus Velocity curve.

Figure 14.3.2.2.3 presents an aircraft Steady State and Instantaneous Turn Rate curve.

Figure 14.3.2.2.4 presents an aircraft Altitude versus Non-Augmented and Augmented Rate of Climb curve.

Figure 14.3.2.2.5 presents an aircraft Altitude versus Velocity curve, for maximum aircraft power.

Figure 14.3.2.2.6 presents an aircraft Altitude versus Velocity curve, for non-augmented aircraft power.

Figure 14.3.2.2.7 presents an aircraft Turn Rate Energy Management versus Velocity curve, for an altitude of 5,000 feet.

Figure 14.3.2.2.8 presents an aircraft Turn Rate Energy Management versus Velocity curve, for an altitude of 20,000 feet.

Figure 14.3.2.2.9 presents an aircraft Turn Rate Energy Management versus Velocity curve, for an altitude of 35,000 feet.

Figure 14.3.2.2.10 presents an aircraft Fuel Flow versus Velocity curve, for maximum aircraft power.

Figure 14.3.2.2.11 presents an aircraft Fuel Flow versus Velocity curve, for maximum non-augmented aircraft Power.

Each of these curves is utilized to provide the reference data for the Independent Verification and Validation (IV&V) tests.

TANN-EM

NATIONAL AIR INTELLIGENCE CENTER

ENGINEERING REPORT

ANALYSIS

This product was prepared by the National Air Intelligence Center (NAIC). It has not been reviewed or approved by the Assistant Chief of Staff, Intelligence, USAF. It may not represent previously approved and coordinated Department of the Air Force, or Department of Defense positions.

COPY # _____

Figure 14.3.2.2.1 NAIC Engineering Report Cover

Flight Simulator Performance

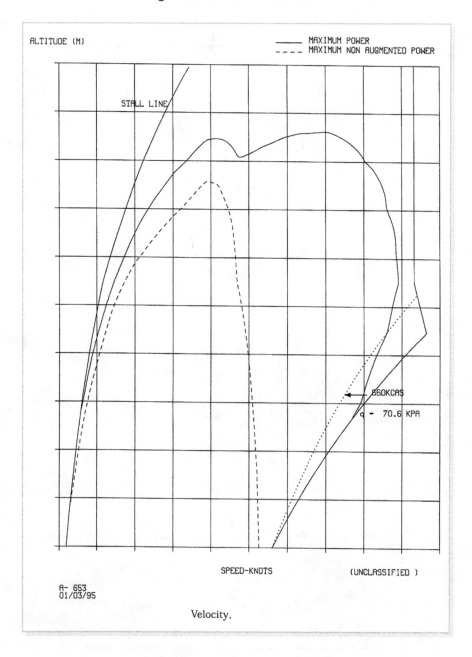

Figure 14.3.2.2.2 Aircraft Altitude versus Velocity curve.

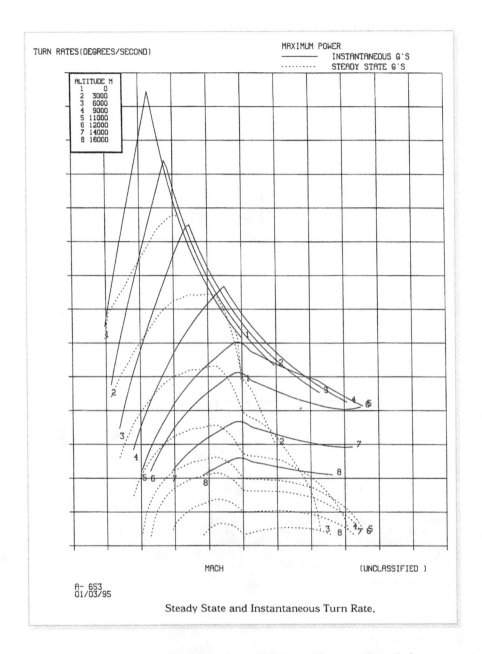

Figure 14.3.2.2.3 Aircraft Steady State and Instantaneous Turn Rate curve

Flight Simulator Performance

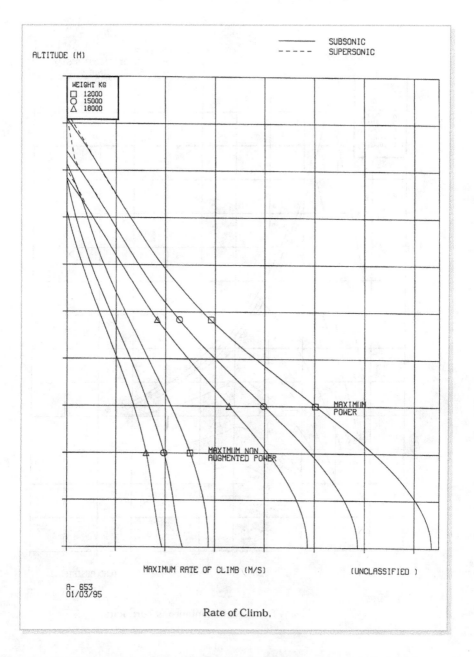

Figure 14.3.2.2.4 Aircraft Altitude versus Non-Augmented and Augmented Rate of Climb curve.

442

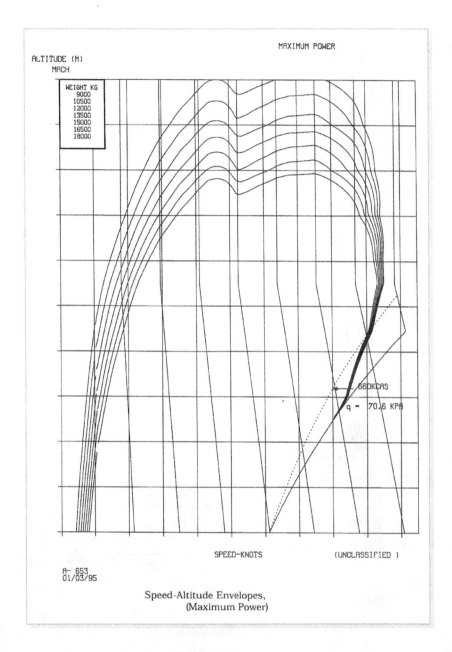

Speed-Altitude Envelopes,
(Maximum Power)

**Figure 14.3.2.2.5 Aircraft Altitude versus Velocity curve,
for maximum aircraft power.**

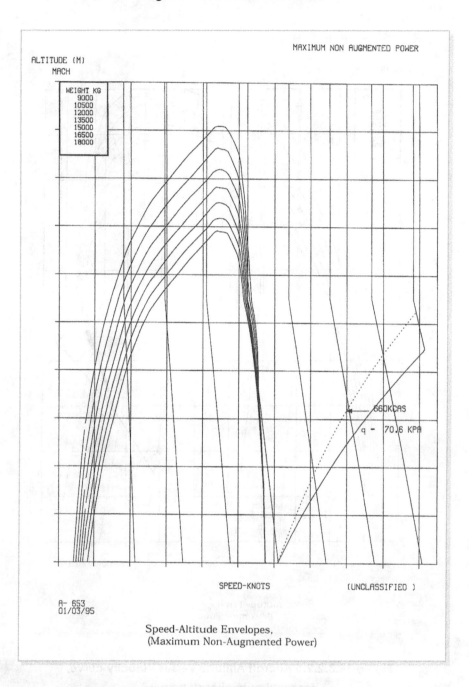

Figure 14.3.2.2.6 Aircraft Altitude versus Velocity curve, for non-augmented aircraft power.

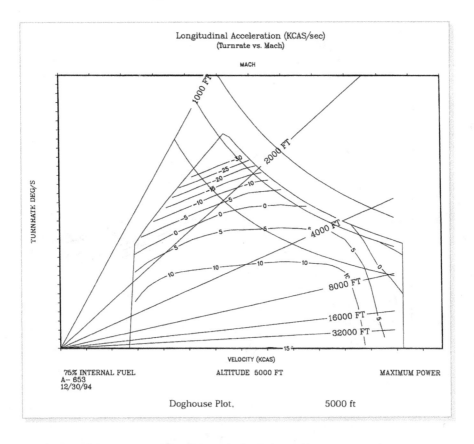

Figure 14.3.2.2.7 Aircraft Turn Rate Energy Management versus Velocity curve, for an altitude of 5,000 feet.

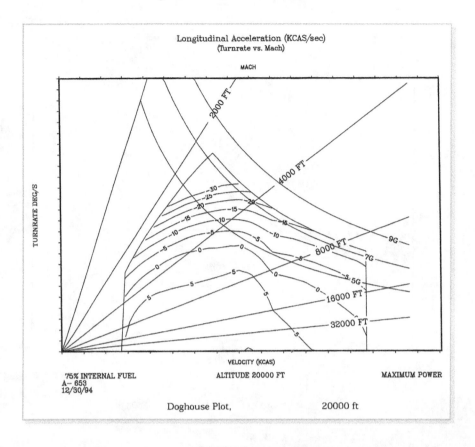

Figure 14.3.2.2.8 Aircraft Turn Rate Energy Management versus Velocity curve, for an altitude of 20,000 feet.

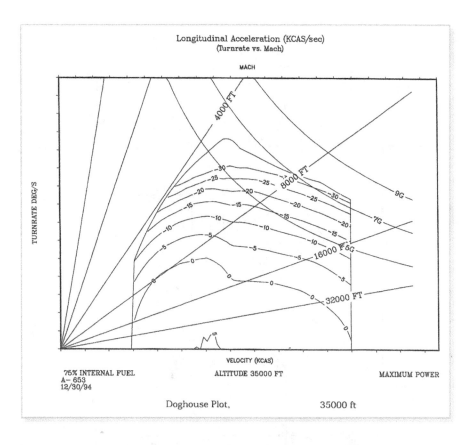

Figure 14.3.2.2.9 Aircraft Turn Rate Energy Management versus Velocity curve, for an altitude of 35,000 feet.

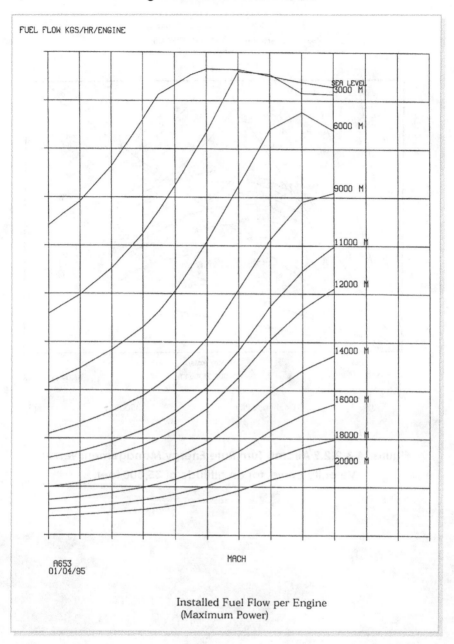

FUEL FLOW KGS/HR/ENGINE

SEA LEVEL
3000 M
6000 M
9000 M
11000 M
12000 M
14000 M
16000 M
18000 M
20000 M

MACH

A653
01/04/95

Installed Fuel Flow per Engine
(Maximum Power)

**Figure 14.3.2.2.10 Aircraft Fuel Flow versus Velocity
curve, for maximum aircraft power**

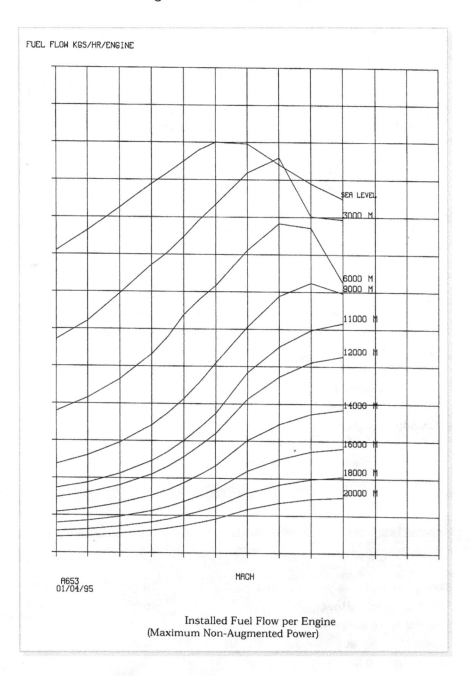

FUEL FLOW KGS/HR/ENGINE

SEA LEVEL

3000 M

6000 M
9000 M

11000 M

12000 M

14000 M

16000 M

18000 M

20000 M

MACH

A653
01/04/95

Installed Fuel Flow per Engine
(Maximum Non-Augmented Power)

**Figure 14.3.2.2.11 Aircraft Fuel Flow versus Velocity curve,
for maximum non-augmented aircraft power**

14.3.2.3 Generation of NGTS Measured Values

The methodology of obtaining the values that will be measured against reference values varies with each flight simulation system. In general the parameters that need to be obtained include the following.

General Parameters

Time

Aircraft Parameters Parameters	Weapon (Missile & Bomb) Detection System) Parameters	Sensor (Radar & Missile
X-Coordinate Position	X-Coordinate Position	Sensor Type
Y-Coordinate Position	Y-Coordinate Position	Sensor Mode
Z-Coordinate Altitude	Z-Coordinate Altitude	Status Flags (On / Off)
Heading (Degrees)	Weapon Type	Bearing (Degrees)
Velocity (Ft/sec)	Launch Flag	
Roll Angle (Degrees)	Launcher ID	
Pitch Angle (Degrees)	Target ID	
Yaw Angle (Degrees)	Hit / Miss	
Turn Radius (Nmi)	Time of Flight (Sec)	
	Velocity of Impact (Ft / sec)	
	Detonation Position	

Aircraft Relational	Weapon Relational Parameters	Sensor Relational Parameters
Range (Nmi)	Range (Nmi)	Range at Detection (Nmi)
Closing Velocity (Ft/sec)		Range at Acquition (Nmi)
Aspect Angle (Degrees)		Range at Tracking (Nmi)
Delta Altitude (Feet)		Range at Missile Fire (Nmi)
Delta Heading (Degrees)		

Some of these parameters need to be documented during each performance evaluation test.

The F-14B/A & D Trainer system allowed the documentation of 100 parameters during each training exercise. The parameters could be documented at a 1 Hz rate up to a 100 Hz rate. Each training exercise was provided a parameter data set that documented the parameters needed to evaluate the model performance.

The NGTS did not have the same capability, as the F-14 B/A & D Trainers, of specifically documenting the parameters. The NGTS utilizes a High Level Architecture (HLA) capability, which allows several tactical environments to be operated together. The HLA system has a Parameter Listener capability that allows the documentation of many of the parameters needed to evaluate the model performance. The HLA Parameter Listener system does not record any relational parameters, such as range, closing velocity and aspect angle.

Presented below, is the methodology utilized to evaluate the NGTS aircraft, weapon, and radar models performance.

The NGTS Instructor Operator Station (IOS) is illustrated below in Figure 14.3.2.3.1

The IOS has a number of Pull Down Menu that is used to operate the NGTS Tactical Environment. Presented below are some of the Pull Down Menu's that facilitate the IOS operation.

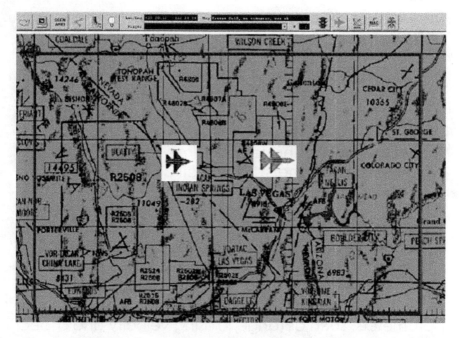

Figure 14.3.2.3.1 NGTS Instructor Operator Station (IOS)

The NGTS Instructor Operating Station (IOS) has the capability of selecting any aircraft, in the NGTS data base, and the aircraft velocity, altitude and heading desired is incorporated, as illustrated in Figure 14.3.2.3.2. The aircraft weapons loadout and radar is put into the aircraft during the model development phase of the program.

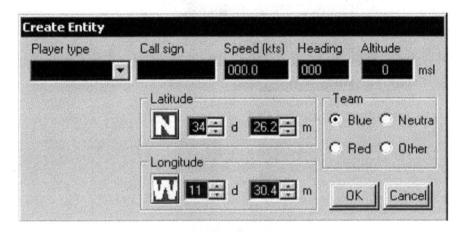

Figure 14.3.2.3.2 Create Aircraft Entity

The Aircraft Console Panel, illustrated in Figure 14.3.2.3.3, provides aircraft details, such as the weapons available.

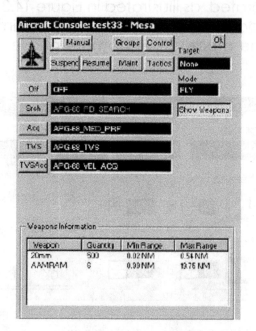

Fire Button

Figure 14.3.2.3.3 Aircraft Console Panel

The aircraft turn rate is generated by flying the aircraft at a certain heading, altitude and speed, and then timing how long it takes to turn to another heading. This is accomplished using the Pull-Down Menu illustrated in Figure 14.3.2.3.4

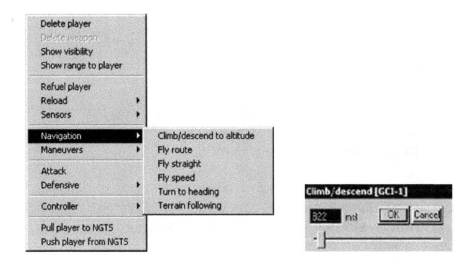

Figure 14.3.2.3.4 Aircraft Parameter Change Menu

The aircraft turn radius is obtained by measuring the range from the turn aircraft, after it turned 180 degrees, to another aircraft that is placed at the point of the start of the turn. The range is obtained from the Pull Down Menu, illustrated below in Figure 14.3.2.3.5

Figure 14.3.2.3.5 IOS Parameter Pull Down Menu

The aircraft climb rate is generated by flying the aircraft at a certain altitude, heading and speed, and then timing how long it takes to fly to a new altitude.

The aircraft acceleration is generated by flying the aircraft at a certain speed, heading and altitude, and then timing how long it takes to fly to a new speed.

The aircraft minimum and maximum velocity is obtained by simply flying the aircraft at various minimum and maximum velocities at various altitudes until the minimum and maximum velocities are found.

The Weapons Time of Flight and Velocity of Impact values are obtained by creating a Shooter Aircraft and a Target Aircraft at a specific heading, altitude, and speed. The missile is fired from a button on the IOS Aircraft Console Panel, shown in Figure 14.3.2.3.3, during the exercise.

The Radar Detection, Acquisition, Tracking, and Missile Fire Ranges are obtained by creating a Shooter Aircraft and a Target Aircraft at a specific heading, altitude and speed. The IOS Aircraft Console Panel, shown in Figure 14.3.2.3.3, identifies the Radar Detection, Acquisition, Tracking and Missile Fire mode and the range is then measured.

For security reasons the specific aircraft, weapons and radar systems are numbered instead of specifically identified. The Rosetta Stone is locked in a safe at work.

14.3.2.4 NGTS Aircraft Validation Test Results

14.3.2.4.1 Aircraft Turn Rate Validation Test Results

Table 14.3.2.4.1.1 below provides the aircraft turn rates for Aircraft Number 1, 2 and 3, at various speeds and altitudes.

Table 14.3.2.4.1.1 NGTS Aircraft Turn Rate Validation Test Results							
(Run May 06 With FOM 1.2.4)							
Speed	Altitude	Aircraft No 1 Ref Turn Rate	Aircraft No 1 Mea Turn Rate	Aircraft No 2 Ref Turn Rate	Aircraft No 2 Mea Turn Rate	Aircraft No 3 Ref Turn Rate	Aircraft No 3 Mea Turn Rate
Knots	Feet	Deg / Sec	Deg / Sec	Deg / Sec	Deg / Sec	Deg / Sec	Deg / Sec
265	5000	8.3 *	10.3	14.0	10.7	14.2	10.7
530	5000	9.7	5.2	16.0	5.6	16.4	Does not turn
410	20000	5.5	6.6	9.6	7.2	10.0	6.8
475	20000	5.7	5.8	.6	6.2	10.0	6.2
550	20000	5.8	Does not turn	9.6	5.2	10.0	Does not turn
600	20000	5.5	4.8	6.6	5.1	8.4	4.75

* Reference values are found in NAIC Engineering Reports, TANN-EM-XX-XXX, Turn Rate Curves.

The NGTS aircraft turn rates are fairly close to the real world reference values except for the 5000 ft altitude at a fairly high speed. The turn rates vary with altitude and with different aircraft.

Change Request: Check aircraft 1, 2 and 3 turn rates at 5000 ft with a 530 knots speed against reference.

14.3.2.4.2 Aircraft Turn Rate "G" Loading

Table 14.3.2.4.2.1 provides the aircraft "G" Force Turn Rate for Aircraft Number 1, 2 and 3, at various speeds and altitudes. The aircraft "G" Force is determined from the equations presented below.

$$Ar = \frac{Vt^2}{R}$$

Where,

Ar = Radial Acceleration

Vt is the aircraft tangential velocity (ft / sec)

R is the aircraft radius during a turn (ft)

Vt (Knots x 1.688) = ft / sec

R (NMI x 6076) = ft

| Alt | Speed | Speed | A | A/C No 1 | B | A/C No 1 | A/C No 2 | B | A/C No 2 | A/C No 3 | B | A/C No 3 |
| | | Speed Kts x 1.688 | Speed² | Turn Radius | A/C No 1 nmi x 6076 | G-Load A / B / 32 | Turn Radius | A/C No 1 nmi x 6076 | G-Load A / B / 32 | Turn Radius | A/C No 1 nmi x 6076 | G-Load A / B / 32 |
Feet	Kts	Ft / Sec	Ft² / Sec²	nmi	Ft	G	nmi	Ft	G	nmi	Ft	G
5000	265	447	199809	.8	4860	1.28	.7	4253	1.46	.7	4253	1.46
5000	530	895	801025	3.0	18228	1.37	2.9	17620	1.42	Does not turn	N/C	N/C
20000	410	692	478864	1.8	10936	1.36	1.7	10329	1.44	1.7	10329	1.44
20000	475	802	643204	2.4	14582	1.37	2.3	13975	1.43	2.3	13974	1.43
20000	550	928	861184	Does not turn	N/C	N/C	4.0	24304	1.1	Does not turn	N/C	N/C
20000	600	1012	1024144	3.8	23088	1.38	3.7	22481	1.42	3.7	22481	1.42
20000	800	1350	1823040	6.7	40709	1.39	6.9	41924	1.35	6.9	41924	1.35

Table 14.3.2.4.2.1 NGTS Aircraft G Force Turn Rate Validation Test Results (Run May 06 With FOM 1.2.4)

The reference result for this test is to ensure that the Turn Rate G-Load does not exceed the maximum allowable G-Load of the aircraft.

The NGTS aircraft turn rate G-Load appears to be about 1.5 G for each aircraft turn. The designer has the aircraft model flying a large turn radius so that the aircraft stays within the 1.5 G-Load. This value is well within the maximum allowable G-Load for the aircraft, but these are not the true values for an aircraft flying in a combat situation. An aircraft probably will sometimes turn in a 1500 ft radius with a 6 G-load during an air-to-air combat engagement.

Change Request: Provide realistic turn rate radius and G-Loading values for an aircraft in a typical mission.

14.3.2.4.3 Aircraft Energy Management Validation Test Results

Table 14.3.2.4.3.1 below provides the aircraft energy management turn rate for Aircraft 1, 2 and 3 at various speeds and altitudes.

		Table 14.3.2.4.3.1 NGTS Aircraft Energy Management Validation Test Results (Run May 06 With FOM 1.2.4)					
Speed	Altitude	Aircraft No 1 Ref Turn Rate @ Zero Gain / Loss	Aircraft No 1 Mea Turn Rate	Aircraft Ref Turn Rate @ Zero Gain / Loss	Aircraft No 2 Mea Turn Rate	Aircraft No 3 Ref Turn Rate @ Zero Gain / Loss	Aircraft No 3 Mea Turn Rate
Knots	Feet	Deg / Sec	Deg / Sec	Deg / Sec	Deg / Sec	Deg / Sec	Deg / Sec
200	5000	7 *	10.4	12	13.8	13	11.4
300	5000	8	9.0	14	9.5	15	9.5
400	5000	10	7.1	16	7.2	16.5	6.5
500	5000	10	5.7	15	5.8	16	5.9
200	15000	4	6.8	7	13.8	11	7.9
300	15000	7	9.0	9	9.4	12	9.6
400	15000	8	7.0	10	6.6	13	7.3
500	15000	7	5.7	7	5.8	12	5.9
200	20000	3	5.4	7	12.5	7	5.4
300	20000	7	9.2	9	9.3	9	9.0
400	20000	8	7.5	10	7.4	0	7.0
500	20000	10	5.7	7	5.9	7	5.7
200	25000	2	3.8	4	10.9	7	4.3
300	25000	4	7.7	5	9.6	9	9.5
400	25000	5	7.0	4	7.3	9	7.5
500	25000	4	5.7	3	5.9	7	5.9
200	35000	-	Does not turn	3.5	6.8	3.5	3.1
300	35000	3		5	9.4	5	6.2
400	35000	5	4.7	3.5	7.3	3.5	7.3
500	35000	7	7.4 / 5.7	3	5.9	3	5.8

* The reference values are obtained from the NAIC Engineering Report, TANN-EM-XX-XXX, Doghouse Plots, for 5000 ft, 20,000 ft and 35000 ft. The reference values for the 15,000 ft and 25,000 ft are obtained from the Close In Combat Threat Model Validation Report, McDonnell Douglas Corporation. The values are taken off the zero energy curves for velocities of 200, 300, 400 and 500 knots.

The NGTS aircraft turn rates appear to be close to values that provide zero energy gain / loss. The aircraft turn rates need to be variable, so that during a maneuver or tactic the aircraft can gain or lose energy, depending on the operation goal. The NGTS aircraft turn rates vary with altitude and aircraft.

Change Request: Provide the aircraft models with the capability of a variable turn rate.

14.3.2.4.4 Aircraft Climb Validation Test Results

Table 14.3.2.4.4.1 below provides the aircraft climb rate for Aircraft 1, 2 and 3 at various speeds and altitudes.

Table 14.3.2.4.4.1 NGTS Aircraft Climb Rate Validation Test Results (Run May 06 With FOM 1.2.4)

A/C No 1, 2 &3 Altitude Feet	A/C No 1 Speed Knots	A/C No 1 Ref Climb Rate Ft / Sec	A/C No 1 Mea Climb Rate Ft / Sec	A/C No 2 Speed Knots	A/C No 2 Ref Climb Rate Ft / Sec	A/C No 2 Mea Climb Rate Ft / Sec	A/C No 3 Speed Knots	A/C No 3 Ref Climb Rate Ft / Sec	A/C No 3 Mea Climb Rate Ft / Sec
10,000 to 20,000	610 Military	125 *	257	610 Military	246	257	600 Military	166	257
10,000 to 20,000	750 Aft Burn	466	317	850 Aft Burn	590	360	920 Aft Burn	583	389
20,000 to 30,000	590 Military	83	264	600 Military	196	255	590 Military	83	267
20,000 to 30,000	900 Aft Burn	291	396	876 Max Aft Burn	426	364	954 Max Aft Burn	383	406

* The reference values are found in NAIC Engineering Reports, TANN-EM-XX-XXX, Climb Rate Curves.

The NGTS aircraft climb rate varies with velocity, but not with altitude, as per the reference values.

Change Request: The aircraft models should provide a variable climb rate, in accordance with real world capabilities.

14.3.2.4..5 Aircraft Maximum / Minimum Speed Validation Test Results

Table 14.3.2.4.5.1 below provides the aircraft minimum and maximum speeds for Aircraft 1, 2 and 3 at various altitudes.

Table 14.3.2.4.5.1 NGTS Aircraft Minimum And Maximum Speed Vs Altitude Validation Test Results (Run May 06 With FOM 1.2.4)						
Aircraft No 1, 2, 3 Altitude	Aircraft No 1 Ref Min Speed	Aircraft No 1 Mea Min Speed	Aircraft No 2 Ref Min Speed	Aircraft No 2 Mea Min Speed	Aircraft No 3 Ref Min Speed	Aircraft No 3 Mea Min Speed
Feet	Knots	Knots	Knots	Knots	Knots	Knots
5000	175 *	217	135	148	125	158
15000	200	217	150	148	150	158
20000	225	217	175	148	175	158
35000	350	217	230	148	250	158
55000	700	217	480	148	900	158
Aircraft No 1, 2, 3 Altitude	Aircraft No 1 Ref Max Speed	Aircraft No 1 Mea Max Speed	Aircraft No 2 Ref Max Speed	Aircraft No 2 Mea Max Speed	Aircraft No 3 Ref Max Speed	Aircraft No 3 Mea Max Speed
Feet	Knots	Knots	Knots	Knots	Knots	Knots
5000	775 *	982	720	876	775	957
15000	925	982	840	876	850	957
20000	1000	982	900	876	925	957
35000	1350	982	990	876	1110	957
55000	1350	982	983	876	1000	957

* The reference values are obtained in the NAIC Engineering Report, TANN-EM-XX-XXX, Velocity Curves, for 5000 ft, 15,000 ft, 20,000 ft, 35,000 ft and 55,000 ft altitudes.

The NGTS aircraft models provide a fixed minimum and maximum speed, independent of altitude. This is a compromise that uses a good part of the velocity reference curves, but does not exactly follow all of the curve values. The NGTS aircraft model minimum and maximum velocity do vary with aircraft.

Presented below is the Figure 14.3.2.4.5.1 Aircraft Altitude versus Velocity curve with the typical measured constant minimum velocity and maximum velocity at any altitude.

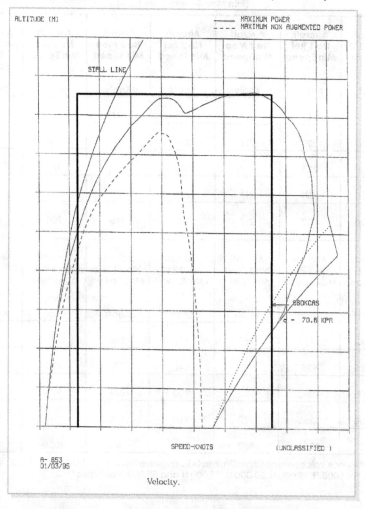

Figure 14.3.2.4.5.1 Altitude versus Velocity Curve

14.3.2.4.6 Aircraft Acceleration Tests

Table 14.3.2.4.6.1 below provides the aircraft accelerations for Aircraft 1, 2 and 3 at various altitudes.

Table 14.3.2.4.6.1 NGTS Aircraft Acceleration Validation Test Results (Run May 06 With FOM 1.2.4)							
Aircraft No	Aircraft Altitude	Aircraft Ref Military Acceleration	Aircraft Ref Aft Burner Acceleration	Aircraft Mea Velocity To Velocity	Aircraft Mea Velocity To Velocity	Aircraft Mea Time	Aircraft Mea Acceleration
	Feet	Ft/Sec²	Ft/Sec²	Knots	Ft/Sec	Sec	Ft/Sec²
1	20,000	8.8 *	18.6	614 to 983	1036 to 1659	39.92	15.6
2	20,000	12.0	24.0	614 to 785	1036 to 1325	19.30	15.0
3	20,000	13.0	23.0	614 to 927	1036 to 1564	34.15	15.5

* The reference values for the aircraft military acceleration and the after burner acceleration came from Al Piranian's (NAWCADPAX Modeling Engineer) Hand Written Document.

The NGTS aircraft model acceleration is constant between the real world values of the non-augmented acceleration and the afterburner acceleration. The NGTS aircraft model accelerations do not vary with aircraft.

Change Request: Provide an aircraft model variable acceleration to include cruise, military and afterburner accelerations.

14.3.2.5 NGTS Weapon Test Results

14.3.2.5.1 Air-To-Air Weapons Test

Table 14.3.2.5.1.1 below provides the air-to air weapon test initial conditions for weapons, randomly selected from section 12.4.3.2

Table 14.3.2.5.1.1 NGTS Air-To-Air Weapon Performance Test Initial Conditions (Run May 06 With FOM 1.2.4)							
Weapons	Orientation	Shooter Aircraft			Target Aircraft		
		X nmi	Altitude Feet	Velocity Knots	X nmi	Altitude Feet	Velocity Knots
1	Nose to Nose	0	20,000	590	15.0	20,000	590
2	Nose to Tail	0	20,000	590	2.5	20,000	590
3	Nose to Nose	0	20,000	590	10.0	20,000	-590
4	Nose to Gnd	0	20,000	590	20	0	20
5	Nose to Tail	0	20,000	590	2.5	20,000	590
6	Nose to Tail	0	20,000	590	2.0	20,000	590
7	Nose to Tail	0	20,000	590	2.5	20,000	590
8	Nose to Tail	0	20,000	590	2.5	20,000	590
9	Nose to Tail	0	20,000	590	2.5	20,000	590
10	Nose to Tail	0	20,000	590	2.5	20,000	590
11	Nose to Nose	0	20,000	590	10.0	20,000	-590
12	Nose to Nose	0	20,000	590	10.0	20,000	-590
13	Nose to Nose	0	20,000	590	10.0	20,000	-590
14	Nose to Tail	0	20,000	590	2.5	20,000	590
15	Nose to Nose	0	20,000	590	10.0	20,000	-590
16	Nose to Gnd	0	20,000	590	20.0	0	20
17	Nose to Gnd	0	20,000	590	20.0	0	20

* The initial conditions are obtained from the F-14 Trainer Test Procedures and Results Report (TTPRR), Grumman Aerospace Corporation.

For security reasons the weapon test initial conditions, in Table 14.3.2.5.1.1, provide weapon numbers instead of names.

In addition, for security reasons the weapon test results presented in pages 472 to 488 provide weapon numbers completely different that those presented in Table 14.3.2.5.1.1. For example the number 1 missile in Table 14.3.2.5.1.1, is not the number 1 weapon in the test results. The Rosetta Stone is locked in a safe at the Pax River Naval Base.

14.3.2.5.2 Surface-To-Air Weapons Validation Test Results

Table 14.3.2.5.2.1 below provides the surface-to air weapon test initial conditions for weapons, randomly selected from section 12.4.3.2

Table 14.3.2.5.2.1 NGTS Surface-To-Air Weapon Performance Tests Initial Conditions (Run May 06 With FOM 1.2.4) *							
Weapon	Orientation	Shooter Aircraft			Target Aircraft		
		X nmi	Altitude Feet	Velocity Knots	X nmi	Altitude Feet	Velocity Knots
18	Nose to Gnd	0	0	0	15.0	20,000	350
19	Nose to Gnd	0	0	0	8.2	4,000	400
20	Nose to Gnd	0	0	0	15.0	20,000	350
21	Nose to Gnd	0	0	0	15.0	20,000	350
22	Nose to Gnd	0	0	0	3.5	2,000	400
23	Nose to Gnd	0	0	0	15.0	20,000	350
24	Nose to Gnd	0	0	0	.9	4,000	400
25	Nose to Gnd	0	0	0	TBD	TBD	TBD
26	Nose to Gnd	0	0	0	TBD	TBD	TBD
27	Nose to Gnd	0	0	0	8.2	4,000	400
28	Nose to Gnd	0	0	0	15.0	20,000	350
29	Nose to Gnd	0	0	0	.5	4,000	400

30	Nose to Gnd	0	0	0	.9	4,000	400
31	Nose to Gnd	0	0	0	15.0	20,000	350
32	Nose to Gnd	0	0	0	3.3	4,000	400
33	Nose to Gnd	0	0	0	TBD	TBD	TBD
34	TBD	TBD	TBD	TBD	TBD	TBD	TBD

* The initial conditions are obtained from the F-14 Trainer Test Procedures and Results Report (TTPRR), Grumman Aerospace Corporation.

For security reasons the weapon test initial conditions, in Table 14.3.2.5.2.1, provide weapon numbers instead of names.

In addition, for security reasons the weapon test results presented in pages 472 to 488 provide weapon numbers completely different that those presented in Table 14.3.2.5.2.1. For example the number 1 missile in Table 14.3.2.5.2.1, is not the number 1 weapon in the test results. The Rosetta Stone is locked in a safe at the Pax River Naval Base.

14.3.2.5.3 Weapons Validation Test Results

Table 14.3.2.5.3.1 below provides the weapons Time of Flight and Velocity of Impact test results.

	Table 14.3.2.5.3.1 NGTS Weapon Performance Test Results (Run May 06 With FOM 1.2.4)				
Weapon Number	Weapon Time of flight (TOF) Ref Sec	Weapon Time of Flight (TOF) Mea Sec	Weapon Velocity Of Impact Ref Ft/Sec	Weapon Velocity Of Impact Mea Ft/Sec	Pass / Failed / Defer Comments
1	9.5	9.0	1347 *	1283	Pass
2	20.5	23.5	4953 *	4858	Pass
3	7.0	3.4	1730 *	1613	Failed: (TOF Low)
4	7.5	5.4	1461 *	1060	Pass
5	TBD	-	TBD	-	Not Tested: (No NGTS Model Available)
6	TBD	18.66	TBD	2447	Not Tested: (Need Reference Data)
7	11.7	12.5	1724 *	1655	Pass: (Full Velocity vs. Time Curve Matches Fairly Well)
8	21.0	22.4	4984 *	2350	Failed: (Velocity of Impact is low)

9	19.5	18.6	1717 *	1499	Pass: (Full Velocity vs. Time Curve Matches Fairly Well)
10	TBD	-	TBD	-	Not Tested: (No Test Available)
11	TBD	-	TBD	-	Not Tested: (No Test Available)
12	8.5	3.5	1902 *	1688	Failed: (Time of Flight and Velocity of impact Low)
13	21.0	24.8	4983 *	3376	Failed: (Time of Flight high and Velocity of Impact low)
14	11.0	10.4	2100 **	2123	Pass
15	32.0	31.9	2296 *	1000	Failed: (Velocity of Impact Low) (Full Velocity vs. Time curve matches somewhat well)
16	26.5	29.1	2950 **	3011	Pass: (Full Velocity vs. Time curve matches somewhat well)
17	16.0	17.9	2500 **	2755	Pass
18	34.0	36.3	2783 *	2363	Pass: (Full Velocity vs. Time curve does not match very well)

19	TBD	-	TBD	-	Not Tested: (No NGTS Model Available)
20	16.0	21.4	1550 **	1025	Failed: (Time of Flight is high and Velocity of Impact Low) (Full Velocity vs. Time curve does not match very well)
21	14.0	14.9	3100 **	3192	Pass: (Full Velocity vs. Time Curve Matches Well)
22	14.5	18.9	1900 **	1804	Failed: (Time of Fight High) (Full Velocity vs. Time Curve Matches Fairly Well)
23	11.5	11.0	2859 *	3021	Pass
24	13.2	11.5	1536 *	2734	Failed: (Velocity of Impact High)
25	11.7	10.7	2012 *	2459	Failed: (Velocity of Impact high)
26	11.4	10.3	2117 *	2700	Pass
27	12.9	10.0	1667 *	1271	Pass
28	11.3	12.5	1827 *	2194	Pass
29	32.0	36.0	1850 **	1500	Failed: (Missile fell far short of target) (Full Velocity vs. Time curve does not match)

30	16.5	16.5	2600**	2550	Pass: (Full Velocity vs. Time Curve Matches Well)
31	11.7	12.6	1724 *	1755	Pass: (Full Velocity vs. Time Curve Matches Well)
32	30.3	28.0	1484 *	1622	Pass: (Full Velocity vs. Time Curve Matches Fairly Well)
33	TBD	-	TBD	-	Not Tested: (No NGTS Model Available)
34	TBD	-	TBD.	-	Not Tested: (No Test Available)

* The reference values are obtained from the F-14 Trainer Test Procedures and Results Report (TTPRR), Grumman Aerospace Corporation.

** The reference values are obtained from the TACTS Range / TRAPS curves, used to compare the full Velocity versus Time curves.

14.3.2.6 NGTS Weapon Curves Validation Test Results

The test results, provided in Table 14.3.2.5.3.1, include 17 weapons passing their test, 10 weapons failing their test and 7 weapons not tested.

The test results provided in Table 14.3.2.5.3.1 are an indication of the accuracy of the weapon at one velocity versus time situation; however it would be a better validation test if the entire weapon velocity versus time curve was generated and compared with its reference curve. This was accomplished for a number of weapons and the data is provided below, in Figures 14.3.2.6.1 to

12. The NGTS CVS Listener was turned on and a weapon was fired from an NGTS aircraft flying at a speed of 590 knots and an altitude of 20,000 ft. The velocity versus time data was collected and the data was plotted against the reference curves.

Weapon Number 7, 9, 15, 21, 30, and 31 are very accurate. Weapon Number 16, 18, 20, 22, and 32 are fairly accurate. Weapon Number 29 needs re-design.

Change Request: Weapon Number 29 needs to be re-designed

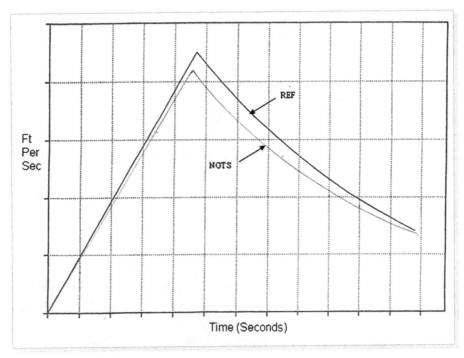

Figure 14.3.2.6.1 Weapon Number 7 Velocity versus Time Curve

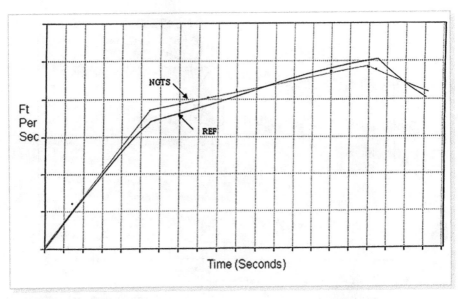

Figure 14.3.2.6.2 Weapon Number 9 Velocity versus Time Curve

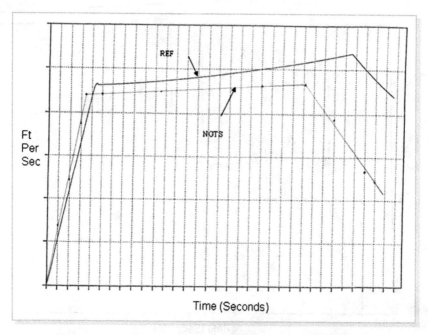

Figure 14.3.2.6.3 Weapon Number 15 Velocity versus Time Curve

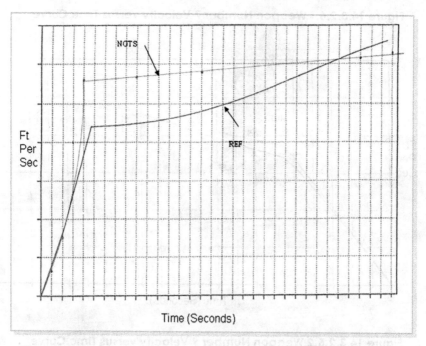

Figure 14.3.2.6.4 Weapon Number 16 Velocity versus Time Curve

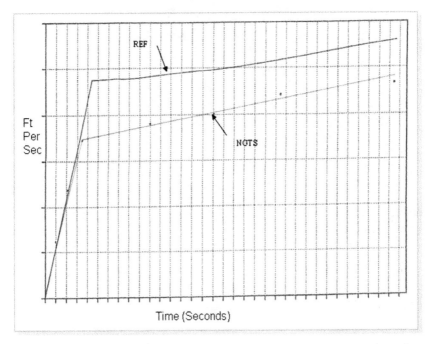

Figure 14.3.2.6.5 Weapon Number 18 Velocity versus Time Curve

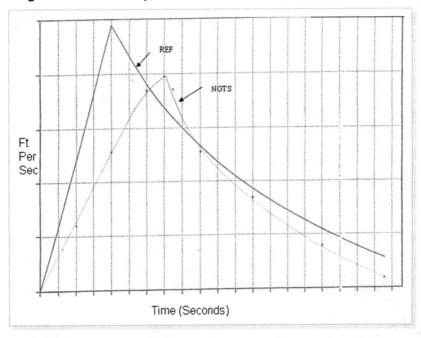

Figure 14.3.2.6.6 Weapon Number 20 Velocity versus Time Curve

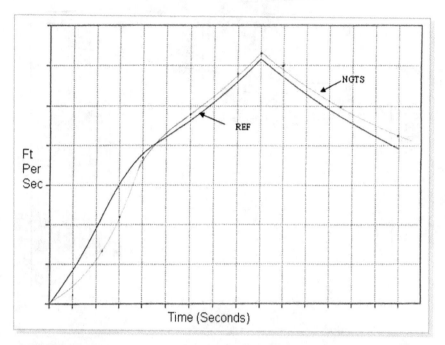

Figure 14.3.2.6.7 Weapon Number 21 Velocity versus Time Curve

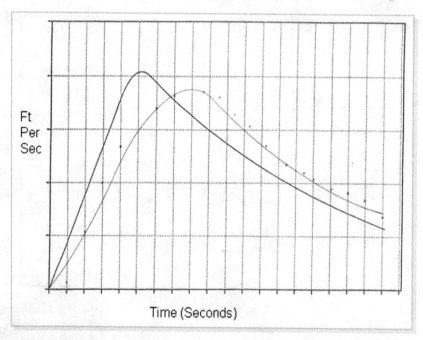

Figure 14.3.2.6.8 Weapon Number 22 Velocity versus Time Curve

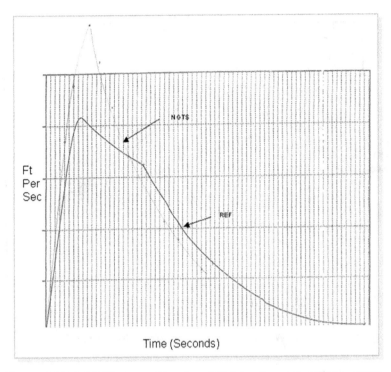

Figure 14.3.2.6.9 Weapon Number 29 Velocity versus Time curve

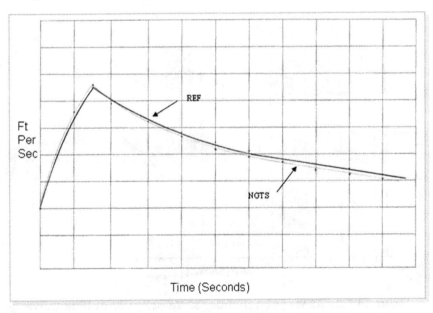

Figure 14.3.2.6.10 Weapon Number 30 Velocity versus Time Curve

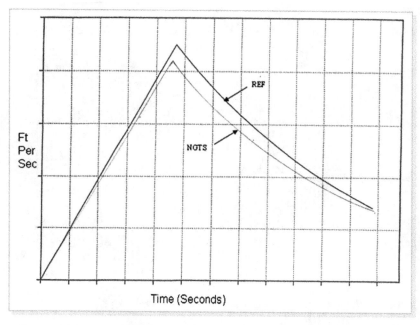

Figure 14.3.2.6.11 Weapon Number 31 Velocity versus Time

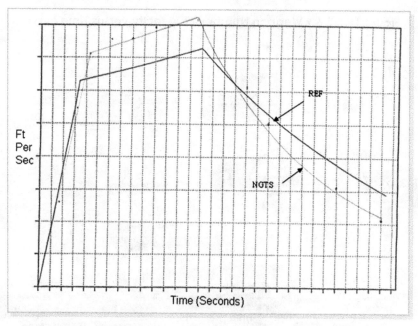

Figure 14.3.2.6.12 Weapon Number 32 Velocity versus Time Curve

14.3.2.7 NGTS Radar Validation Test Results

14.3.2.7.1 Aircraft Air-To-Air Radar Validation Test Results

Table 14.3.2.7.1.1 below provides the aircraft radar detection, acquisition, tracking and fire ranges.

Table 14.3.2.7.1.1 Air-To-Air Validation Test Results (Run May 06 With FOM 1.2.4)

Radar Number and Orientation	Shooter	Aircraft	Target	Aircraft	Test Results							
	Alt Feet	Vel Kts	Alt Feet	Vel Kts	Ref Detect Range nmi	Mea Detect Range nmi	Ref Acq Range nmi	Mea Acq Range nmi	Ref Track Range nmi	Mea Track Range nmi	Ref Fire Range nmi	Mea Fire Range nmi
1 NOSE TO TAIL	20K	590	20K	300	38.0 *	Starts in Search	38.0	47.4	29.0	47.3	Pilot Fires	8.8
1 NOSE TO NOSE	20K	590	20K	-590	38.0	Starts in Search	38.0	46.2	29.0	45.3	Pilot Fires	Aircraft did not fire ***
2 NOSE TO TAIL	20K	590	20K	300	80.0	Starts in Search	80.0	39.1	40.0	29.9	Pilot Fires	Aircraft did not fire

(Continued)

2 NOSE TO NOSE	20K	590	20K	-590	80.0	Starts in Search	80.0	36.7	40.0	30.5	Pilot Fires	Aircraft did not fire ***
3 NOSE TO TAIL	20K	590	20K	300	38.0	Starts in Search	38.0	39.7	26.6	**	Pilot Fires	Aircraft did not fire
3 NOSE TO NOSE	20K	590	20K	-590	38.0	Starts in Search	38.0	39.1	26.6	**	Pilot Fires	Aircraft did not fire

* The reference values were obtained from the STILO Office Reports
** Radar went back and forth from the Search mode to the Acquire mode to the Search mode.
At a 23 nmi range the aircraft did a Drag maneuver.
*** The aircraft did a Drag maneuver

The NGTS Air-to-Air Radar Models Number 1, 2 and 3, did not exactly match the reference data for the acquisition range or tracking range. Also, the Aircraft Number 1, 2 and 3, in most cases did not automatically fire on the Target Aircraft, as required. Activating the fire button did cause a weapon fire.

Change Request: Review Air-to-Air Radar Model Number 1, 2, and 3 designs to improve the radar performance.

For security reasons the specific radar systems are numbered instead of specifically being identified. The Rosetta Stone is locked in a safe at work.

14.3.2.7.2 Aircraft Surface-To-Air Radar Validation Test Results

Table 14.3.2.7.2.1 below provides the SAM Site radar detection, acquisition, tracking and firing ranges.

Table 14.3.2.7.2.1 Surface-To-Air Validation Test Results (Run May 06 With FOM 1.2.4)

Radar Number And Orientation	Shooter Aircraft Alt (Feet)	Shooter Aircraft Vel (Kts)	Target Alt (Feet)	Aircraft Vel (Kts)	Test Results Ref Detect Range (nmi)	Mea Detect Range (nmi)	Ref Acq Range (nmi)	Mea Acq Range (nmi)	Ref Track Range (nmi)	Mea Track Range (nmi)	Ref Fire Range (nmi)	Mea Fire Range (nmi)
1 Gnd To Air	0	0	20K	300	75.6 *	Starts in Search Mode	75.6	95.2	49.0	92.6	Operator Fire	44.3 Miss 37.4 Hit
2 Gnd To Air	0	0	20K	300	49.0	Starts in Search Mode	49.0	76.6	22.0	38.2	Operator Fire	33.5 Miss Short **

3 Gnd To Air	0	0	20K	300	27.0	Starts in Search Mode	27.0	13.6	13.5	13.3	Operator Fire	4.4 Hit
4 Gnd To Air	0	0	20K	300	39.0	Starts in Search Mode	39.0	30.3	33.0	30.0	Operator Fire	11.2 Hit

* The reference values were obtained from the STILO Office Reports

** Activation of fire button only allowed missile to fly a short range.

The NGTS Surface-to-Air Radar Models Number 1, 2, and 3, did not exactly match the reference data for the acquisition range or tracking range. The NGTS Aircraft Number 1, 2, 3 and 4, in most cases, did automatically fire on the target aircraft, as required. In addition, activating the fire button did cause a weapon fire.

Change Request: Review Surface-To-Air Radar Model Number 1, 2, 3 and 4 designs to improve the radar performance.

14.3.3 Flight Simulator Tactical Environment System Evaluation

The Flight Simulator tactical environment system evaluation pursues a number of mission scenarios that includes the aircraft models, weapon models, radar models and electronic warfare models within each mission scenario. The mission scenarios include the following,

- Air-to-Air Intercept
- Air-to-Ground Weapon Delivery
- Air-to-Ship Weapon Delivery
- Integrated Air Defense System (IADS) / (SAM Sites)
- Electronic Warfare (EW)
- Outer-Air-Battle
- Air Combat Maneuvering (ACM) / (Dog Fight)

The Flight Simulator tactical environment system evaluation will pursue the evaluation of the F-14 B/A & D Trainer Flight Simulators because I have more data on these evaluations.

The F-14 B/A & D Trainer provides the ability to record 100 parameters at rates of up to 100 Hz, during the mission scenario. The evaluation procedure first consists of generating the initial conditions of the mission scenario.

A typical Air Combat Maneuvering (ACM) mission scenario initial conditions document is presented below in Figure 14.3.3.1

Initial Conditions:

Ownship:

Position:
  Latitude: 36 ° N
  Longitude: 78 ° W
Altitude (Ft):   15K
Heading (Deg):   90
Velocity (Ft/Deg):   350
Status: Non-Killable

Five Threats:

Type:	MIG-31	MIG-31	MIG-31	SU-24	SU-24
Skill Level:	4	4	4	4	4
Position					
Latitude:	36 ° N	36 ° 03' N	36 ° 06' N	36 ° N	36 ° 03' N
Longitude:	76 ° 30' W	76 ° 30' W	76 ° 30' W	77 ° W	77 ° W
Altitude (ft):	18K	18K	18K	8K	8K
Heading (Deg):	270	270	270	270	270
Velocity (Ft/Sec):	350	350	350	350	350

Figure 14.3.3.1 Air Combat Maneuvering (ACM) Mission Scenario Initial Conditions

The Air Combat Maneuvering (ACM) Initial Conditions is a typical example of mission scenario initial conditions. Each of the other mission scenarios has a set of initial conditions that is specifically generated for that specific exercise.

The next step is to generate a list of parameters that will be collected during the mission scenario exercise.

The Parameter Data Set is presented below, in Figure 14.3.3.2

Parameter Type	No of Parameters
Time:	1
Ownship Data:	
Coordinates (X, Y, & Z)	3

Heading	1
Velocity	1
Threat Fighter Data:	
Coordinates (X, Y, Z)	15
Heading	5
Velocity	5
Roll Angle	5
Pitch Angle	5
Yaw Angle	5
Range	5
Weapon Data:	
Launch Flag	5
Launcher ID	5
Target ID	5
Weapon Type	5
Range	5
Hit / Miss	5
Reason For Miss	5
Time Of Flight	5

Total	91

Figure 14.3.3.2 Air Combat Maneuvering (ACM) Mission Scenario Data Set

The Air Combat Maneuvering (ACM) Data Set is a typical example of a mission scenario data set. Each of the other mission scenarios has a data set that is specifically generated for that specific exercise.

The Air Combat Maneuvering (ACM) mission scenario exercise was conducted and the graphical test results are presented below in Figure 14.3.4.3. The exercise details the position of the F-14 Aircraft ownship flown by a pilot and the position of the five threat aircraft, during the exercise. Two threat aircrafts leave the air combat combat zone during the exercise, which is caused by the tactics rules applied to the specific aircraft.

Analysis of the test results data provides information regarding the threat aircraft model performance.

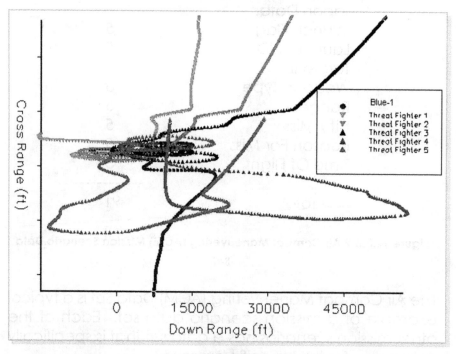

Figure 14.3.3.3 Air Combat Maneuvering (ACM) Mission Exercise Graphics

The weapon firing test results of the mission exercise is presented below in Table 14.3.3.1. The table of test results includes the launcher, target, hit / miss results, launch range, time-of-flight and the missile velocity. Analysis of the test results data provides information regarding the weapon model performance.

Table 14.3.3.1 Air Combat Maneuvering (ACM) Mission Exercise Test Results								
Launch Time	Impact Time	Launcher	Target	Hit/ Miss (Reason For Miss)	Launch Range	Time- Of - Flight	Missile Travel Distance	Missile Average Velocity
(Sec)	(Sec)				(Ft)	(Sec)	(Ft)	(Ft / Sec)
64.1	132.3	Blue 1	Threat 5	Hit	161,000	66.8	106,400	1592
115.7	139.5	Blue 1	Threat 5	Hit	70,000	23.6	50,650	2146
180.8	243.9	Blue 1	Threat 3	Miss (18)	129,000	63.3	88,100	1391
190.1	245.9	Threat 1	Blue 1	Miss (5)	115,000	55.3	85,460	1545
192.2	24.7	Threat 3	Blue 1	Hit	114,000	45.5	93,500	2054
192.2	249.1	Threat 2	Blue 1	Miss (5)	113,000	55.7	86,000	1543
214.7	249.1	Threat 1	Blue 1	Miss (?)	81,800	32.0	69,100	2159
248.0	264.5	Threat 1	Blue 1	Hit	44,300	16.5	34,000	2060
250.1	266.6	Threat 2	Blue 1	Hit	44,100	16.0	34,400	2150

14.3.4 Tactical Environment Rules Evaluation

The evaluation of the flight simulator models is designated as Independent Verification and Validation (IV&V). Verification is a measurement against the flight simulator design. Validation is a measurement against the systems real world performance. Three flight simulator elements that are designs that must use verification to evaluate the performance are the following.

- Tactics
- Visual Priority
- Skill Level

A set of rules are generated for the aircraft tactics, visual priority and the training skill level. In general these rules are generated by aircraft crewstation experts. The first issues is that you have to know the rules before you can evaluate the elements, and the contractors did not want to provide the rules. At one meeting I asked for the tactics rules and a contractor said, "Why do you need the tactics rules?" During our evaluation I tried to get the pilots to evaluate the threat tactics and they could not provide me their subjective evaluation. I then paid the contractor $75,000 to document the high fidelity aircraft rules.

Detailed below are a couple of the tactics rules, the visual priority rules and the Skill Level Rules and the evaluation of the rules.

14.3.4.1 Tactic Rules and the Rules Evaluation

The F-14D Trainer threat aircraft tactic rules identify a mission situation and then perform a typical maneuver.

The maneuver options include the following for the threat aircraft.

Threat is greater than 15 miles from ownship

A-Pole	Glib
D-Cross Over (26L, 26R, 26LS, 26RS)	Gun Jink
D-Lead Around (11L, 11R, 11LS, 11RS)	Hook (15, 20, 25)
F-Pole, F-Pole1	Lead Around (24, 29)
Lookup (FH, RH)	Pincer (25, 30, 35)
Barrel Roll (L & R)	Sweep (15, 15S, 17L, 17R, 20R
Def. Break	20S, 23L, 23R, 25R, 25S)
Break Away (15, 20, 25)	Trail (1.5, 3)
Drag (15, 20, 25)	

Threat is less than 15 miles from ownship

Basic	Aileron Roll
Circling	Barrel Roll
X, Y, Z Location	Head On
Spiraling	Jimking
Pure Pursuit	Station Keeping
Bulls Eye	Dive
Split	Max G Pull Up
Immelman	Yo Yo
Cuban Eight	Climbing Turn
Loop	

Presented below is a portion of the high fidelity threat aircraft rules when the threat aircraft is less than 15 miles from the ownship. The tactics rules include skill level rules.

1. IF: The threat is approaching the ownship from the rear and the closing speed is greater than 200 kts.
   THEN: Skill Level 4 expects a High Yo Yo tactic (This slows down the closing speed)
   Skill Level 1 expects a Lead Pursuit tactic

2. IF: The threat is approaching the ownship from the rear and the closing speed is less than 25 kts.
   THEN: Skill Level 4 expects a Low Yo Yo tactic (This speeds up the closing speed)
   Skill Level 1 expects a lead pursuit tactic

3. IF: The ownship is approaching threat from the rear.
   THEN: Skill Level 4 and 1 expects a Climbing Hard Turn tactic

The IF situational is setup during a tactical exercise. Presented below is the THEN graphics presentation of the tactical exercise. If the graphic presentation follows the THEN rule then the rule is verified.

Presented in Figure 14.3.4.1.1 is a graphical representation of the Rule Number 3, which agrees quite well with the rule. Several other tactics rules were evaluated and they also agreed quite well.

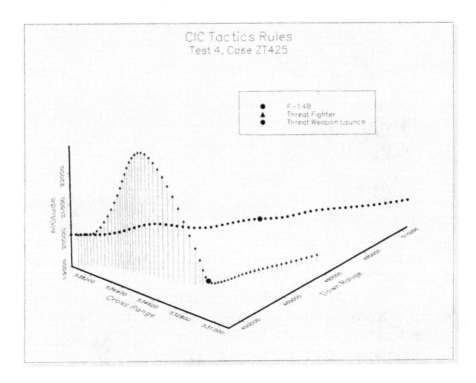

Figure 14.3.4.1.1 Tactical Rule Number 3 Graphical Representation

14.3.4.2 Visual Display Tactical Environment Rules and Rules Evaluation

The visual display tactical environment rules depend on what the aircraft crewpersons think is the most important elements to be presented during a mission exercise. The visual display display rules depend on the number of tactical environment display projectors provided in the flight simulator facility and the fidelity of each projector.

Presented below are the visual display rules for the F-14 B/A Trainer Facility.

F-14 B/A Trainer Visual Display Rules

Object	Visual Range (nm)	Visual Priority Number
Aircraft Carrier	15	10,000
Threat Aircraft	12	840
Missiles	3	80
Airfields	10	9,500
Ground Targets	20	1,350
Surface Targets	20	1,600
Own-Ship Weapons	3	300
Weather	60	300
Wingman	12	10,000

The F-14 B/A Trainer Flight Simulator tactical environment visual display system includes five high fidelity target projectors, five medium fidelity inset projectors, and five low fidelity background projectors.

An F-14 B/A Trainer tactical environment visual display mission exercise initial conditions is presented in Figure 14.3.4.2.1

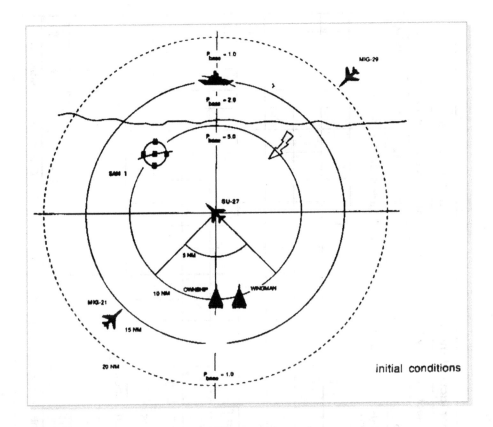

**Figure 14.3.4.2.1 F-14 B/A Trainer Tactical Environment Visual Display
Exercise Initial Conditions**

Presented in Table 14.3.4.2.1, are the test results of the F-14 B/A Trainer tactical environment visual display presentation.

Time (sec)	High Fidelity Target Projectors					Medium Fidelity Target Projectors					Low Fidelity Background Projectors				
	1	2	3	4	5	6	7	8	9	10	11	12	13	14	15
0	Threat 1	Threat 2	Threat 3	Threat 4	Threat 5	Wing-man	None	Carrier	Carrier	Carrier	Carrier	Carrier	None	None	None
10	→					→									
20	Wing-man														
30						Wing-man									
40	Wing-man					Threat 1									
50	Wing-man					Threat 1									
60	Threat 1					None									
70						None	None								
80						None	None						None		
90						Chaff	Chaff						Chaff		
100						Chaff	Chaff						None		
110						None	None						None		
120	Threat 1	Threat 2	Threat 3	Threat 4	Threat 5	None	None	Carrier	Carrier	Carrier	Carrier	Carrier	None	None	None

Table 14.3.4.2.1 F-14 B/A Trainer Tactical Environment Visual Display Test Results

130 to 180 No Change Occurred

	Threat 1	Threat 2	Threat 3	Threat 4	Threat 5	None	None	None	Carrier	Carrier	Carrier	Carrier	Carrier	Carrier	None	None	None
190	→	→	→	→	→				→	→	→	→	→	→	→	→	→
200						None											
210						Missile	Missile										
220						Chaff	Chaff										
230						Chaff	Chaff										
240						Missile	None	→									
250						Missile		None									
260						Missile		Chaff									
270						Chaff		None									
280						None		→									
290						None			→	→	→	→	→	→			
300	→	→	→	→	→	None			Carrier	Carrier	Carrier	Carrier	Carrier	Carrier	None	None	None

The test results detailed in Table 14.3.4.2.1 do not exactly agree with the visual display priority rules, but they do provide test results on what actually happens during a mission scenario. The military can decide if this is the visual display priority that they want or do they want any changes in the rules.

Presented below are the tactical environment visual display priority rules for the F-14 D Trainer Facility. Obviously, the F-14D Trainer tactical environment visual display priority rules are very different than the F-14 B/A Trainer rules. The F-14 D Trainer rules were generated by McDonnell Douglas Corporation, with the help of the F-14D Aircraft crewpersons. The F-14 B/A Trainer rules were generated by Grumman Aerospace Corporation, with the help of the F-14 B/A Aircraft crewpersons.

Aircraft

Closure Rate (0-6)
Range (0-12)
Friend or Foe (2-3)
Radar Designated (1-1.5)
Time Displayed (1-2)

Missile

Closure Rate ()-6)
Range (0-12)
Radar Designate (1-1.5)
Locked On (1-2)
Boost (1-2)
Time Displayed (1-2)

14.3.4.3 Skill Level Rules and Rules Evaluation

The F-14 B/A & D Trainer Skill Level rules were not totally provided although they were requested many times. Presented below are a couple of the high level skill level rules that were provided.

Skill Level 4 has perfect situational awareness.
Skill Level 1 knows less than exact information, such as the ownship position, altitude or energy state.
Lower skill levels tend to fire weapons earlier.
Higher skill levels wait for better opportunities

The Skill Level rules evaluation consisted of a couple of pilots flying an Air Combat Maneuvering mission scenario with different skill levels made functional. Their subjective opinion indicated that a Skill Level 4 threat was definitely more aggressive. The Skill Level rules were never fully evaluated.

14.4 Flight Simulator Aircraft Engine Performance

One of the most important flight simulation systems is the simulation of the ownship aircraft being provided.

This book provides the F-18 and F-14B/A & D Aircraft Trainers. The evaluation of the Trainer Flight Simulator crewstation operation is provided by a combination of the pilots that fly the aircraft and a team of crewstation experts that make specific measurements on various elements of aircraft dynamics. My observations was that the pilots subjective aircraft dynamic feelings and the specific crewstation experts measurements did not always agree, but the joint evaluation process provided the compromises that made the final crewstation dynamic adjustments fairly good.

14.4.1 Aircraft Engine Performance Specifications

Presented below are the F-14B/A Trainer crewstation dynamic performance specifications, taken from the Systems Requirements Document (SRD), by the Naval Air Warfare Center, Training System Division, Orlando, Florida. The SRD numbering system is maintained below.

3.1.5.1.11 F-14A and F-14B Flight Simulation: The F-14B/A Flight Simulation shall provide simulation of the dynamics of the F-14A and F-14B aircraft based on the solution of the six degrees-of-freedom equations of motion. The simulation shall provide the responses of aerodynamic, powerplant, weapon system and environmental parameters in accordance with aircraft data. The simulation shall include aerodynamic, inertial and aero elastic effects for all altitudes simulated. Static and dynamic flight characteristics shall be simulated over the full flight envelope of the aircraft.

3.1.5.1.11.1 Aerodynamics

3.1.5.1.11.1.1 Description: The aerodynamic simulation shall simulate the aerodynamic spectrum of the F-14B and F-14A Aircraft for,

a. Taxi.
b. Takeoff.
c. Climb.
d. Descent.
e. Approach.
f. Landing.
g. Aerodynamic Ground Effects.
h. Target Turbulence.
i. Cruise from sea level to absolute ceiling.
j. Maneuvering aerodynamics for maximum airframe G, high alpha sustains flight, inverted flight, departure, spin entry, and spin recovery.
k. Incremental aerodynamics due to malfunctioned or degraded systems or components, including flight controls and propulsion.
l. Incremental aerodynamics effects due to external stores both symmetrical and asymmetrical.
m. Direct Lift Control (DLC).
n. Aircraft Carrier turbulence, entitled burble.
o. Environmental components.
p. Jetwash.
q. Effects due to the Reduced Arrestment Thrust System.
r. Mid-compressor bypass.
s. Glove Vane.

3.1.5.1.11.2 Mass and Moment of Inertia, and Center of Gravity

3.1.5.1.11.2.1 Description: The mass, moment of inertia and center of gravity simulation incorporates the physical characteristics of the F-14A and F-14B Aircraft configurations, fuel tank locations, and external stores. The three qualities, mass, moment of inertia, and center of gravity are time dependent and functions of aircraft loading and configurations.

The simulation of the mass, moment of inertia and center of gravity shall compute the effects due to the following.

a. Fuel load distribution and consumption for the fuel burn curve.
b. Wing sweep, including position and rate.
c. External stores, including weapons, pods and tanks.
d. Landing gear motion.
e. Wing configuration changes.
f. Icing.

3.1.5.2 Malfunctions: System malfunctions shall be simulated and modeled from aircraft data and simulate the aircraft characteristics in the same manner as in the aircraft. The malfunctions and failures shall cause primary system and related supporting system indications to be simulated as in the design basis aircraft. Engine and fuel system malfunctions shall be implemented separately for the left and right engines. Provisions to combine malfunctions shall be provided except for conditions which are mutually exclusive. Displays and indicators at both the trainee station and IOS shall respond to the inserted malfunctions.

3.1.5.3 Tolerances: The tolerances shall apply throughout the range of operation of the design basis aircraft regardless of whether the range is normal or abnormal. The tolerance shall apply to the following design basis aircraft data,

a. Directly measured aircraft flight test data.
b. Corrected and normalized flight test data.
c. Data extracted from test stand and wind tunnel tests.
d. Analytical data.

Unless otherwise specified, the tolerance figures are plus and minus values. The tolerances shall be those at the place the values are read at the computer, IOS and trainee station. In cases where the accuracy of the operational aircraft instrument and indicator is less than the tolerance specified, the operational aircraft instrument accuracy shall apply.

3.1.5.3.1 General: Simulated aircraft performance characteristics not covered by specific tolerances shall be assigned the following general tolerances within which the trainer shall operate.

a. Total Mass	1 %
b. Moment of Inertia	1 % or 0.1 % of maximum values whichever is greater.
c. Attitude Accuracy	1 degree
d. Attitude Resolution	0.1 degree
e. Center of Gravity Position	0.2 % of mean aerodynamic chord.
f. Other	5 %

3.1.5.3.2 Curve Slope: The curve slope of a trainer performance curve shall be within + and − 10 % of the curve slope of the corresponding trainer criteria curve at points where the criteria curve is continuous.

3.1.5.3.3 Control System Tolerance: Force and moment characteristics shall be measured at the pilot's nominal point of application. Control stick and rudder pedal positions shall be defined as the position of those mechanical elements in contact with the pilot.

a.	Surface Deflections vs. Control	0.5 degrees deflection
b.	Pitch Control Forces vs. Control Deflection	1 lb or 10 %, whichever is greater
c.	Roll Control Force vs. Control Deflection	1 lb or 10 %, whichever is greater
d.	Pedal Force vs. Deflection	2 lbs or 10%, whichever is greater
e.	Breakout plus friction force	1 / 2 lb
f.	Time for operation of trim system The flap actuation, speed brake actuation	5 %
g.	Damping Ratio	0.05
h.	Damped Natural Frequency	10 %
i.	Amplitude Response	10 %

3.1.5.3.4 Flying Qualities Tolerances:

Steady state trim points are presented below:

a.	Angle of Attack vs. Trim Airspeed	0.5 unit angle-of-attack
b.	Control Deflection vs. Trim Airspeed	10 % or 1 degree, whichever is greater
c.	Trim Surface Deflection Indication vs. Airspeed	10 % or 1 degree, whichever is greater
d.	Approach Speed vs. Gross Weight	1 Knot IAS

Longitudinal trim changes due to thrust changes and activation of appurtenanances, such as landing gear, wing flaps, speed brakes, wing sweep:

a.	Control Position Change	0.5 degrees

b. Control Force Change   1 lb or 10 % whichever is greater.
c. Pitch Angle Change   1 degree
d. Angle of Attack Change   1 degree
e. Altitude Change   Greater of 10 ft or 10 % of total change in altitude.
f. Airspeed vs. Change   Greater of 5 knots Indicated airspeed (KIAS) or 10 % of the total change in airspeed.

Static Longitudinal Stability:

a. Control Deflection vs. Airspeed   0.5 deg
b. Surface Deflection vs. Airspeed   0.5 deg or 10 %, whichever is greater.
c. Control Force vs. Airspeed   0.5 lb or 10 %, whichever is greater.
d. Angle of Attack vs. Airspeed   0.5 unit angle-of-attack.

Dynamic Longitudinal Stability:

a. Damping Ratio   0.05
b. Undamped Natural Frequency   10 %
c. Amplitude Response   10 %

Maneuvering Longitudinal Stability:

a. Stick Force per Unit Acceleration   10 % or 1 lb.
b. Control Deflection vs. Acceleration   0.5 degree or 10 %, whichever is greater.
c. Angle-of-Attack vs. Acceleration   0.5 degree or 10 %, whichever is greater.
d. Surface Deflection vs. Acceleration   0.5 degree or 10 %, whichever is greater.

Static Lateral Directional Stability:

a.	Lateral Control Deflection vs. Sideslip Angle	0.5 % or 10 %, whichever is greater.
b.	Lateral Control Force vs. Sideslip Angle	1 lb or 10 %, whichever is greater.
c.	Lateral Surface Deflections vs. Slideslip Angle	0.5 deg or 10 %, whichever is greater.
d.	Roll Angle vs. Sideslip Angle	1.0 deg or 10 %, whichever is greater.
e.	Directional Control Deflection vs. Sideslip Angle	0.5 deg or 10 %, whichever is greater.
f.	Directional Control Force vs. Sideslip Angle	0.5 deg or 10 %, whichever is greater.

Dynamic Lateral Directional Stability:

a.	Dutch Roll Period vs. Airspeed	10 %
b.	Dutch Roll Damping vs. Airspeed	0.05
c.	Roll to Sideslip Ratio	10 %
d.	Sideslip Angle vs. Time	10 % or 1 deg of peak amplitude, whichever is greater.

Spiral Stability:

a.	Roll Angle vs. Time	20 % and convergent or divergent as in aircraft.

Lateral Control Effectiveness:

a.	Roll Angle vs. Time	10 %
b.	Sideslip Angle vs. Time	10 %
c.	Roll Rate vs. Time	10 %
d.	Roll Mode Time Constant	25 %

Engine out Flying Qualities:

a. Minimum Control Airspeed   3 knots
b. Dynamic Response to Sudden   15 %
      Engine Failure

Stall Characteristics:

a. Buffet Onset Speed vs. Gross Weight   2 knots
b. Stall Speed vs. Gross Weight   2 knots
c. Buffet Onset Angle of Attack   0.5 Unit
d. Stall Angle of Attack   0.5 Unit

3.1.5.3.5 Power Plant Tolerance:

a. Fuel Flow   5 %
b. Revolutions per minute (RPM)   5 %
c. RPM vs. Power Lever Position   1 % RPM at idle and
                                   greater than
                               90 % RPM, 2 % elsewhere.

d. Engine Windmilling Speed   1 % RPM
e. Engine RPM vs. Time   5 %
f. Exhaust Gas Temperature   3 % from idle to 75 % RPM,
                                   1 % above 75 % RPM
                                   and at idle.

g. Oil Pressure   5 %
h. Thrust   3 %, or 0.3 % of maximum
                                   value whichever is
                                   greater.

i. Light Off Time   10 %
j. Fuel Quality   5 %

3.1.5.3.6 Performance Characteristics:

a. Rate of Climb 5 % or 50 ft per minute whichever is greater

b. Level Acceleration or Deceleration 5 %

c. Maximum Airspeed 3 knots or 1 % whichever is greater

d. Turn Performance 5 % or 0.1 g, whichever is greater

3.1.5.3.7 Ground Handling Characteristics:

a. Heading Angle vs. Time 10 %

b. Heading Angle vs. Time 10 %

c. Ground Speed vs. Time 10 %

3.1.5.3.8 Takeoff and Landing Characteristics:

a. Nosewheel Lift-off Speed vs. Gross Weight and CG Position 2 knots

b. Takroff Airspeed 2 knots

c. Takeoff Time 1 sec

d. Stopping Time 1 sec

e. Rudder and Aileron Effectiveness Speeds 5 knots

f. Distance 10 % of instantaneous aircraft value.

3.1.5.3.9 System Operation Tolerance:

a. Hydraulic Pressure 100 pounds per square inch (psi)

b. Hydraulic Transients 1 sec

c. Electrical Indications 5 %

d. Power, Electrical 5 %

e. Electrical Device Load 5 %

f.	Extension, Retraction Times	15 %
g.	General Time Delay	10 %
h.	Control Force vs. Control Deflection	10 %
i.	Accumulator Pressure	10 %
j.	Actuator Response Time	1 sec
k.	Standby Compass	3 deg
l.	Free Air Temperature	2 deg
m.	Accelerometer	0.1 g
n.	Cockpit Altitude	100 ft
o.	Gyro Horizon Indicator	2 deg
p.	Turn Indicator	1 / 10 needle width
q.	Sideslip	1 / 4 ball width

3.1.5.3.10 Navigation Tolerances:

a.	Relative Bearing	2 deg
b.	Localizer Beam Location	1 ft
c.	Localizer Beam Approach Bearing	0.2 deg
d.	Glidepath Beam Location	10 ft
e.	Glidepath Beam Angle	6 min of arc.
f.	Gyro Precession Rate	25 % or 2 deg per hour, whichever is less.
g.	Magnetic Variation	0.1 deg
h.	Field Elevation	10 ft
i.	Signal Attenuation vs. Distance	25 % maximum range.
j.	Radio Beam Width	20 %
k.	Distance Indicator	0.5 miles
l.	Radio Facility Location	0.1 mile

14.4.2 Aircraft Engine Validation Evaluation

The evaluation of the flight Simulator aircraft aerodynamics is pursued by a team of engineers and pilots that fly the specific aircraft. Measurements are pursued for every specification parameter, presented in Section 14.4.1, and test results analyzed. The evaluation of the F-14A/B & D Aircraft Trainer was conducted by a team headed by Tom Santangelo from the NAS Patuxent River. The aircraft aerodynamics and flight control evaluation was provided during the F-14A/A & D Trainer evaluation cycle. F-14B/A & D Aircraft pilots played an important roll in this evaluation.

In general, the aerodynamics and flight control system trainer evaluations have work reasonably well for most trainers, with one exception that I describe below.

The major experience I specifically had with the evaluation of an aircraft aerodynamics and flight control evaluation occurred on the SH-2F Anti-submarine Helicopter Trainer Program, installed at NAS Norfolk, Virginia and NAS North Island, California. I had a very good aircraft aerodynamic modeling engineer and two very good aircraft flight control system modeling engineers. The SH-2F Flight Simulator Trainer hardware included a beamsplitter / spherical mirror / CRT visual display system, a six degree-of-freedom motion base system and a McFadden flight control system. Each of these hardware systems was fairly good quality state-of art for the 1980's flight simulator period. The modeling engineers could listen to inputs from the helicopter pilots and make changes in the modeling equations that reflected the specific pilot inputs.

Measurements determined that the time delay between the visual and motion cues were up to 1 second time.

The delay time for a typical flight simulator should not be more than 150 ms. The 1 sec time delay simply was a result of the computer and the visual, motion base and flight control systems. The SH-2F Helicopter Trainer Computer was a Sperry 8800 Computer.

A number of SH-2F Helicopter Pilots flew the SH-2F Trainer and provided verbal inputs to the modeling engineers and the modeling equations were changed to satisfy their verbal inputs. After a period of time the pilots were satisfied with the trainer aerodynamics and flight control.

A team of trainer aerodynamics and flight control engineers, from NAS Patuxent River, came to the SH-2F Helicopter Trainer Facility and measured the system against a series of actual SH-2F Helicopter flight measurements that were taken over a period of months. They asked my modeling engineers to change the modeling equations in accordance with their measurements. They also had two SH-2F Helicopter pilots assist with their evaluation.

After the NAS Patuxent River Evaluation Team left the SH-2F Helicopter Facility, the facility pilots attempted to fly the trainer and found that the modeling changes did not allow them to properly fly the helicopter. The modeling equations were changed back to the facility pilot's requests.

The NAS Patuxent River Evaluation Team again traveled to the NAS Norfolk Facility and again requested that the modeling equations be changed to their measurements.

Again, the SH-2F Helicopter Trainer Facility pilot's could not fly the aircraft. As the Program Manager, I was in a major dilemma between the NAS Patuxent River Evaluation Team and the SH-2F Helicopter Trainer Team. I think the problem was partly caused by the 1 second time delay between the visual and motion cues. As program manager, I decided to side with the SH-2F Helicopter Trainer pilot's and change the equations back to their requirements. The NAS Patuxent River Evaluation Team would not sign off on the SH-2F Helicopter Trainer and therefore the NAS Norfolk pilot's could not use the trainer for their required flight time.

14.4.3 Flight Simulator Transport Time Delay

One major problem with the Flight Simulator design is the transport time delay. The transport time delay is the time between a pilots pitching up or down on the stick flight control and the observation of the pitch up or down on the visual display system. The transport time delay is illustrated, for the F-14B/A & D Flight Simulator, in Figure 14.4.3.1

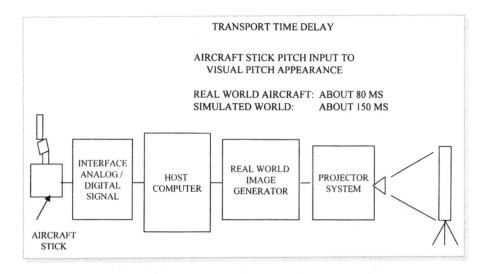

Figure 14.4.3.1 Flight Simulator Transport Time Delay Illustration

The transport time delay for some flight simulators is very large and in some cases more than one second. Flight simulator specifications allow a transport time delay of 150 ms. Coordination of the visual cue to the motion cue in a flight simulation is very important and quite often the difference is substantial. This is a major problem in flight simulation design.

Evaluation of the trainer flight simulator transport time delay is not always an easy procedure. Many years ago I designed and built a lab type SH-2F Helicopter Research Flight Simulator that included an aircraft stick and CRT visual display. I applied a signal generator that simulated a stick input up and down and viewed the horizon on the CRT display move up and then down. I then decreased the signal generator value until the horizon did not move up and down, but stayed stationary at the horizontal position. This was the transport time delay and the value was determined to be about one second in time.

14.5 Flight Simulator Performance Evaluation Stories.

During my 48 years of Department of Defense service I have worked on the design and development of the following Research and Trainer Flight Simulators. The F/A-18 Aircraft Trainer Flight Simulator, the F-14 B/A Aircraft Trainer, the F-14D Aircraft Trainer Flight Simulator, the F-14D Research Flight Simulator, the SH-2F Helicopter Trainer Flight Simulator, the F-14A Aircraft Dynamic Flight Simulator, the Crewstation Design Research Flight Simulator, the SH-2F / SH-60 Helicopter LAMPS Flight Simulator, the F-111 Aircraft Research Flight Simulator, and the CH-53 Helicopter Research Flight Simulator. Each of these Research and Trainer Flight Simulators required hardware, software, modeling, interface, and human factor performance evaluation.

Enclosed are a couple of stories detailing the evaluation of the Research and Trainer Flight Simulators.

1. F-18 Trainer Flight Simulator Tactical Environment [NextGeneration Threat System Aircraft Model] (2003-2006)

The Next Generation Threat System (NGTS) aircraft models were evaluated and found to have no variable turn rate, climb rate or acceleration rate. The software code currently has to be changed to obtain a different value. Developing proper tactics rules requires variable turn rates, climb rates and acceleration rates. The NGTS Program Manager decided to pursue a new aircraft model with these variable capabilities. The NGTS evaluation test results also determined that the aircraft models provided a single minimum and maximum speed at all altitudes.

Systems Engineering provides complete requirement details, including the Trainer Flight Simulator mission requirements, that would alleviate a number of the NGTS limitations.

2. F-14 B/A & D Trainer Flight Simulator (1986-2001)

I was given the opportunity, in 1986, of developing a procedure to conduct verification and validation tests on the F-14D Trainer Flight Simulator. I used the experience I had with modeling the CH-46 Helicopter Inflight Escape system in developing a computer printout of the F-14D Trainer mission exercises and used the printouts to analyze the aircraft, weapons, radar and electronic warfare performance. The F-14D Trainer system provided the capability of printing out 100 parameters, during a training exercise. I generated a series of aircraft mission scenarios that were used to evaluate the trainer performance capability. A data set of mission parameter was collected during the mission exercise and the test results were analyzed to determine the specific performance results. The validation test procedure proved to be very successful in the evaluation of the F-14B/A & D Trainer tactical environment.

Collecting the full 100 parameters during a training exercise at a 100 Hz rate caused the flight simulators computer to slow down to a point that the information on the aircraft crewstation displays is delayed.

Quite often there are side affects to any technical approach that provides a good evaluation of a systems performance.

3. F-14A Aircraft Dynamic Flight Simulator (1982-1985)

The F-14A Aircraft Dynamic Flight Simulator was designed and developed to study the situation awareness of a pilot departing into a spin and the information that the pilot obtains to reduce the affects and recover from the spin. A pilot departing into a spin provided a 7 G force on the pilot, so that the pilot was basically incapacitated. They key element in the evaluation of the F-14A Aircraft Dynamic Flight Simulator was to ensure that the dynamics of the simulator exactly matched an aircraft real world spin. Many pilots were brought in to fly the Dynamic Flight Simulator and they evaluated the spin forces, but of course none of them had ever been in a serious spin in the real world, so their evaluation was somewhat subjective. The pilots used their best judgment in tuning the centrifuge force structure.

A cockpit display was designed to provide aircraft critical spin information, such as the direction of the spin (Clockwise or Counterclockwise) the aircraft spin attitude (Up-Side-Down or Right-Side-Up) and the procedure that a pilot could take to depart from a spin.

A couple of years later a pilot, in California, departed into a serious spin and later said that he thought the spin forces in the F-14A Aircraft Dynamic Flight Simulator were not exactly the same as he felt in the real world. This pilot was a good source for tuning the facility.

4. Crewsystem Design Flight Simulator (1975-1996)

The Crewsystem Design Flight Simulator was designed and developed with a generic crewstation and the facility was used to evaluate new aircraft technology, such as liquid crystal displays, all display cockpits, and battle force mission exercises, using a number of flight simulators. The generic crewstation was designed to replicate a general aircraft crewstation, but was not meant to provide a specific aircraft crewstation. I always thought that if a flight simulator were to be used in training then the crewstation should be an exact replica of the aircraft crewstation. This flight simulator was used to evaluate general functional technology and so a generic crewstation was appropriate.

Each of the aircraft technology issues were basically subjectively evaluated by aircraft experts, but I think that providing a formal experimental design exercise with all new technology developed would better document the performance of the technology.

5. SH-2F / SH-60 Helicopter LAMP Flight Simulator (1970-1990)

The SH-2F / SH-60 Helicopter LAMPS Research Flight Simulator was designed and developed to evaluate the capability of using a helicopter in anti-submarine warfare mission exercises. Initially the system was designed with a SH-2F Helicopter crewstation. The SH-2F crewstaion included a pilot crewstation and an Aircraft Tactical Officer (ATO) crewstation. The program was later incorporated into the SH-60 Helicopter and the crewstation included a pilot crewstation, an Aircraft Tactical Officer (ATO) crewstation and a Sensor Operator (SO).

The SH-2F / SH-60 Helicopter LAMPS Research Flight Simulator was basically evaluated by having a number of anti-submarine crewstation persons perform a number of experimental designs. The experimental designs included the evaluation of a cursor, evaluation of the aircraft keyset, evaluation of the anti-submarine buoy systems, evaluation of the anti-submarine displays, etc. The experimental design exercises not only evaluated aircraft equipment, but also indirectly evaluated the crewsystem performance capability. I was surprised that the quiet crewstation person had the best performance test results and the loud extravert crewstation person had the worst performance test results.

6. CH-53 Helicopter Flight Simulator (1965-1968)

The CH-53 Helicopter Flight Simulator was designed and developed to introduce new technology into the aircraft.

The purpose of the technology was to accommodate the mission operation. The aircraft would be supplied with new display technology, including a vertical image of the world and a map display of the terrain below. In 1968, there was not a formal evaluation of a flight simulator program, so a number of helicopter crewpersons simply flew the CH-53 Helicopter Research Flight Simulator and subjectively provided their evaluation of the system. The new display technology was years ahead of the time it would actually be incorporated into an aircraft, but today some of this technology is standard.

Chapter 15 Glossary, Acronyms and Index

15.1 Glossary

Some of the definitions are taken from the Naval Air Weapon Center, Point Mugu, Handbook.

Accessibility – A measure of the relative ease of admission to the various areas of an item for the purpose of operation or maintenance.

Assembly – A number of parts or subassemblies or any combination thereof jointed together to perform a specific function and capable of disassembly.

Amplifier – An electronic device used to increase signal magnitude or power.

Amplitude Modulation – A method of impressing a message upon a carrier signal by causing the carrier amplitude to vary proportionally to the message waveform.

Angle Jamming – An electronic warfare technique, when azimuth and elevation information from a scanning fire control radar is jammed by transmitting a jamming pulse similar to the radar pulse, but with modulation information out of phase with the returning target angle modulation information.

Antenna Beamwidth – The angle, in degrees, between the half power points of an antenna beam. This angle is also that between the center of the mainlobe and the first null. The angle is given for both horizontal and vertical planes unless the beam is circular.

Aperture – In an antenna, the portion of the plane surface area near the antenna perpendicular to the direction of maximum radiation through which the major portion of the radiation passes. The effective scattering aperture area can be computed for wire antennas which have no obvious physical area.

Asynchronous Pulsed Jamming – An effective form of pulsed jamming. The jammer nearly matches the pulse repetition frequency (PRF) of the radar; then it transmits multiples of the PRF. It is more effective if the jammer pulsewidth is greater than that of the radar. Asynchronous pulsed jamming is similar to synchronous jamming except that the target

lines tend to curve inward or outward slightly and appear fuzzy in the jammed sector of a radar scope.

Barrage Noise Jamming – Noise jamming spread in frequency to deny the use of multiple radar frequencies to effectively deny range information.

Beacon – A system wherein a transponder in a missile receives coded signals from a shipboard radar guidance transmitter and transmits reply signals to a shipboard radar beacon receiver to enable a computer to determine mission position.

Beam – The beam is to the side of an aircraft or ship.

Bingo – The fuel state at which an aircraft must leave the area in order to return and land safely.

Blanking – The process of making a channel or device non-effective for a certain interval. Blanking is used for retrace sweeps on the CRT or to mask unwanted signals such as blanking ones own radar from the onboard Radar Warning Receiver.

Burn-Through Range – The ability of radar to see through jamming. Usually described as the point when the radar's target return is a specified amount stronger that the jamming signals.

Carrier Frequency – The basic radio frequency of the wave upon which modulations are impressed.

Chaff – Ribbon like pieces of metallic materials which are dispensed by aircraft or ships to mask or screen other targets. The radar reflections off of the chaff may cause tracking radar to break lock on the target.

Channel – The communications medium by which modulated electro-magnetic waves convey information between a transmitter and the intended receiver(s). The primary attribute of a channel is its bandwidth, measured in Hertz (Hz).

Clutter – Undesired radar returns resulting from man-made or natural objects, including chaff, sea, ground, and rain, which interface with normal radar system observations.

Collimation – The procedure of aligning fire control radar system antenna axes with optical line of sight, thereby ensuring that the radar will provide for correct target illumination and guidance beam positioning.

Contact – An entity that has been detected and localized by a sensor (usually by means of reflected or self-radiated energy).

Constructive Simulation – A combat simulation that models the states and activity of compound military units (such as, units at echelons above the level of individual personnel, systems, or vehicles/platforms) at discrete time steps, which typically are of the order of minutes or hours of real time.

Core Unit – A conceptual construct to provide a convenient reference mechanism for: (a) the location manned by the program system operator/coordinator, (b) the data processing function that ensures that a consistent simulated combat environment is presented to all participants, and (c) the data archiving function that ensures post-mission availability of a complete consolidated exercise history database.

Cross Pole Jamming – Cross Pole Jamming is a monopulse jamming technique where a cross polarized signal is transmitted to give erroneous angle data to the radar. The component of the jamming signal with the same polarization as the radar must be very small.

Data Transfer – Data transfer refers to a mechanism that exchanges initialization, mission, and post-mission records between each aircraft and the ground subsystem. This mechanism may involve the use of a data storage medium, device, or system, and/or employ a data link for data transfer.

Debriefing Subsystem – The debriefing subsystem is a ground/surface-based station that is able to take individual (participant) recorded data from whatever medium or device data is recorded on and merge it with other mission data to create a debrief capability. The debriefing subsystem has all the displays, functions, and controls to provide a robust graphical debrief.

Decibel (DB) – The dimensionless unit for expressing the ratio of two values of power, current or voltage. The number of decibels being equal to the following equation. $DB = 10 \log P2 / P1$.

Deception – The deliberate radiation, re-radiation, alteration, absorption or reflection of electromagnetic energy in a manner intended to mislead the enemy interpretation or use of information received by his electronic systems.

Designation – The assignment of fire control radar to a specific target by supplying target coordinate data to the radar system.

Detection – The technique of recovering the amplitude modulation signal superimposed on a carrier.

Dish – A microwave reflector used as part of a radar antenna system. The surface is concave and is usually parabolic shaped.

Doppler Effect – The apparent change in frequency of an electromagnetic wave caused by a change in distance between the transmitter and the receiver during transmission and reception.

Duplexer – A switching device used in radar to permit alternate use of the same antenna for both transmitting and receiving.

Duty Cycle – The ratio of average power to peak power, or ratio of pulse length to interpulse period for pulsed transmitter systems. Interpulse period is equal to the reciprocal of the pulse repetition rate.

Effective Radiated Power (ERP) – Input power to the antenna in watts times the gain ratio of the antenna. When expressed in DB, ERP is the transmitter power (Pt) plus the antenna gain (Gt).

Emitter – Any device which emits electromagnetic energy.

Emulation – The imitation of all or part of one system by another, so that the imitating system accepts the same data, executes the same functions, and achieves the same results as the imitated system.

Entity – An object, item, or thing. A simulation of a military platform, system, weapon, sensor, etc. The object may be identified as a simulated object and have the characteristics, function, and behavior of a real instrumentation participant and be integrated into the scenario and processing of the system.

Event Data – A periodic combat system data including initial target detection, target identification, target classification, weapon lock-on, weapon launch, and countermeasure activation/mode change.

Exercise – The execution of one or more related missions.

Exercise Fire – The launching of a weapon that contains an instrumentation/telemetry package in place of the warhead and no explosives other than its propellants.

The weapon is designed to turn away from the target prior to impact point.

Fail-Safe – The design feature of a part, unit, or equipment which allows the item to fail only into a non-hazardous mode.Fault Coverage Factor – The dominant mechanism in fault tolerant systems employing standby redundancy is generally not exhaustion of spares. Rather, system failure is usually due to an inability to perform all of the following functions prior to the time that the level of service.

First Harmonic – The fundamental frequency.

Frequency Agility – A radar ability to change frequency within its opening band, usually on a pulse-to-pulse basis. This is an electronic warfare technique employed to avoid spot jamming and to force the jammer to go into a less effective barrage mode.

Fundamental Frequency – The tuned frequency, carrier frequency, center frequency, output frequency or operating frequency.

Gain – The value of power gain in a given direction relative to an isotropic point source radiating equally in al directions.

Gaming Area – Postulated geographical domain of a system training mission. The gaming area will typically exceed the system connectivity area. The Detached Cluster capability will provide instrumented training for real participants within the gaming area but outside the connectivity area. The system computer-based functions for scenario generation and for exercise information display (such as, geographical and political boundary depiction) may be invoked throughout the gaming area.

Gate (Range) – A signal used to select radar echoes corresponding to a very short range increment. Range is computed by moving the range gate or marker to the target echo; an arrangement which permits radar signals to be received in a small selected fraction of the time period between radar transmitter pulses.

Global Positioning System – The Global Positioning System (GPS) is a system that provides positioning and timing information to military and civilian users worldwide. The GPS has a space segment that consists of a constellation of 24 satellites worldwide that provide distance measurement features for user position definition. A GPS receiver at the user location processes the signals received from a group of the visible satellites to produce user time-space-position information (TSPI).

Ground System – The ground (surface) system is a set of land-based or ship-based equipment necessary to support all or selected elements of training mission including setup, monitor, control and debrief. The ground/surface system consists of commercial personal computer equipment that can be configured to support an easily transportable capability or a fixed training range infrastructure and interface capability. The ground/surface system also includes datalink transfer relays.

Ground Track – The path identified by the vertical projections of a participant's time sequential three-dimensional positions onto the local ground plane.

Harmonic – A sinusoidal component of a periodic wave or quantity having a frequency that is an integral multiple of the fundamental frequency. For example, a component which is twice the fundamental frequency is called the second harmonic.

Hertz – The unit of frequency equal to one cycle per second.

Image Jamming – Jamming at the image frequency of the radar receiver. Barrage jamming is made most effective by generating energy at both the normal operating and image frequency of the radar.

Ingress – Go into the target area.

Interlock – An interlock is an automatic switch which eliminates all power from the equipment when an access door, cover, or plate is removed.

Intermittent Use System – Systems that have relatively long periods of standby or inactivity between uses.

Jamming – The deliberate radiation, or reflection of electromagnetic energy with the object of impairing the use of electronic devices, equipment, or systems by an enemy.

Jink – An aircraft maneuver which sharply changes the instantaneous flight path but maintains the overall route of flight.

Jittered PRF – An anti-jam feature of certain radar systems which varies the PRF consecutively, and randomly, from pulse to pulse to prevent enemy ECM equipment from locking on and synchronizing with the transmitted PRF.

Land-based Threat – A collective term for those categories of "real", such as, surrogate replicas, and simulated weapons (and associated sensor systems [such as, height-finder radar] that have the capability to damage/ destroy an aircraft. This includes anti-aircraft (AAA) guns, small arms barrage fire domes, and surface-to-air missile systems. Although some land-based threats represent weapons that can be installed on mobile ground platforms, they are not operational while in motion.

Line Replaceable Unit (LRU) – Naval Sea Systems Command term for the type of equipment for which organization-level maintenance performs fault isolation and removal/ replacement actions.

Live Fire (surface/air participant context) – The launching/ firing of a real weapon with either an explosive warhead or an instrumentation/telemetry package, in place of the warhead.

Live Fire (subsurface participants) – The launching of a weapon that contains an explosive warhead, or is not designed to turn away from the target prior to impact.

Live Monitor Subsystem – The live monitor subsystem is a ground/surface system that supports live-monitor operations.

Local Processor – The computing resources, hardware and software, embedded within a participant instrumentation package that coordinates the activities of the other onboard elements.

Low Activity Aircraft (LAA) – An aircraft that is within the mission area, but not directly connected within a mission network. LAA are tracked, position only, by Federal Aviation Administration (FAA) or other external sources. The LAA track data is supplied to the system for display in the ground/surface system.

Light Amplification By Stimulated Emission OF Radiation (LASER) – A process of generating coherent light. The process utilizes a natural molecular phenomenon whereby molecules absorb incident electromagnetic energy at specific frequencies, store this energy for short but usable periods, and then release the stored energy in the form of light at particular frequencies in an extremely narrow frequency band.

MACH Number – The ratio of the velocity of a body to the speed of sound in the medium that is being considered in the atmosphere. The speed of sound varies with temperature and atmospheric pressure, therefore so does the Mach number.

Magnetic Anomaly Detector (MAD) – A means of detecting changes in the earth's magnetic field caused by the presence of metal in ships and submarines.

Maintainability – The measure of the ability of an item to be retained or restored to a specified condition when maintenance is performed by personnel having specified skill levels using prescribed procedures and resources, at each prescribed level of maintenance and repair.

Mission – A set of warfare operations executed within the scope of an authored scenario that are directed towards achieving desired training objectives.

Modulation – The process where some characteristic of one wave is varied in accordance with some characteristic of another wave. The basic types of modulation are angle modulation and amplitude modulation.

Modulation, Amplitude – The Amplitude Modulation changes the amplitude of a carrier wave in responses to the amplitude of a modulating wave.

Modulation, Frequency – The Frequency Modulation is varied in proportion to the amplitude of the modulating wave.

Modulation, Phase – The Phase Modulation is varied in proportion to the amplitude of the modulating wave.

Multiplex – Simultaneous transmission of two or more signals on a common carrier wave. The three types of multiplex are called time division, frequency division, and phase division.

Net – A group of intercommunicating subscriber nodes sharing the same channel. Each subscriber node has the potential for connectivity with any other node in the net, but full n by n connectivity (where n is the number of subscriber nodes in the net) is not necessarily achieved at any given instant.

Network – An organizational structure for the collection of communication resources aggregated for convenient reference. A network may be composed of multiple nets and its composition may vary as a function of time.

No Drop Weapon Scoring (NDWS) – No Drop Weapon Scoring (NDWS) provides the ability to simulate air-to-surface non-guided weapons and determine the real-world point of impact and/or splash points. If applicable, the associated scoring/damage to targets within the weapons' lethal distance of the point of impact is provided. NDWS includes bombs, rockets, and mines.

Noise Jamming – A continuous random signal radiated with the objective of concealing the aircraft echo from the enemy radar.

Notch – The portion of the radar velocity display where a target disappears due to being notched out by the zero Doppler filter.

Operator – The person responsible for monitoring and/ or controlling the system instrumentation on a participant.

Organizational Level Maintenance – Maintenance which is the responsibility of, and is performed by, a using organization on its assigned equipment. Its phases normally consist of inspecting, servicing, lubricating, adjusting, and replacing parts, minor assemblies, and subassemblies.

Oscillators – Devices which generate a frequency.

Part – One piece, or two or more pieces joined together which are not normally subject to disassembly without destruction of designed use.

Portable System – A portable system is defined as a system that can be moved from location to location.

Participant – A platform/vehicle or site that is part of the training exercise. Participants may be either instrumented or not.

Primary Network – The collection of communication resources that provide connectivity between the system core instrumentation set and participants.

Propagation – The travel of waves through or along a medium.

Provide – A functional characterization of intent that implies the supplying of something needed which will not be available from another source.

Pseudo – Computer-generated. Not part of the real world. For example, a pseudo track is computer-generated simulation of a contact's position and movement.

Pulsed Doppler – The type of radar that combines the features of pulsed radar and dopper radar.

Pulse Modulation – A special case of amplitude modulation where the carrier wave is varied at a pulsed rate.

Pulse Repetition Frequency (PRF) – The rate of occurrence of a series of pulses, such as 100 pulses per second.

Pulse Repetition Interval (PRI) – Time between the beginning of one pulse and the beginning of the next.

Pulse Width – The interval of time between the leading edge of a pulse and the trailing edge of a pulse.

Radar – Radio detection and ranging.

Radar Cross Section (RCS) – A measure of the radar reflection characteristics of a target. It is equal to the power reflected back to the radar divided by power density of the wave striking the target.

Range – A range is defined as the distance from one system to another system over a specified geographic area.

Range Boundaries – Range Boundaries define the airspace and ground training areas within which missions are executed.

Range Gate – A gate voltage used to select radar echoes from a very short range interval.

Range Gate Pull Off (RGPO) – Deception technique used against pulse tracking radar using range gates. Jammer initially repeats the skin echo with minimum time delay at a high power to capture the circuitry. Time delay is progressively increased forcing the tracking gates to be pulled away from the target echo. Frequency memory loops or transponders provide the variable delay.

Range Rate – The rate at which a radar target is changing its range with respect to the radar.

Radar Resolution – The minimum separation in angle or in range between twp targets which the radar is capable of distinguishing.

Radio Frequency Interference (RFI) – Any induced, radiated or conducted electrical disturbance or transient that causes undesirable responses or malfunctioning in any electrical or electronic equipment.

Rangeless Operation – Rangeless operation occurs when there is no RF connectivity (or datalink) between an air system element and a ground system element.

Realism – The extent to which the operator's experience during simulated operation corresponds to experiences that would have been gained during actual operation of the system under a given set of circumstances.

Real-time – Real-time shall be define as a computer computation that is processed in the same time as the real world time, such as the following.

1. Pertaining to actual time during which physical process or events transpire, (normal clock time-the timing resolution required is relative to the system, task, event, or participant's perspective).

2. Pertaining to the performance of a computer simulation (a) during the actual time that the related physical process transpires, or (b) in which the performance is constrained by time requirements imposed by other processes, in order that results can be used in interaction with other internal or external physical or simulated processes.

3. Pertaining to processes that can be influenced by human intervention while they are in progress.

Real-time Training – Real-time training occurs when the system participants receive nearly immediate feedback concerning their combat maneuvers, weapons employment, and countermeasure employment during a mission and within the scope of training objectives and established tactics.

Reliability – The duration or probability of failure free performance under stated conditions. The probability that an item can perform its intended function for a specified interval under stated conditions. (For Non-redundant items this is equivalent to definition (1). For redundant items this is equivalent to the definition of mission reliability).

Scan – To traverse or sweep a sector or volume of airspace with a recurring pattern, by means of a controlled directional beam from a radar antenna.

Scenario – The mission context consisting of geographical elements, mission elements (participants and potential targets) and their assigned characteristics, rules of engagement, and general planned activity directed towards satisfying desired training objectives for a designated set of participants within a mission. The scenario is generally authored pre-mission, but may be modified during real-time operations.

Scenario Author – The individual responsible for creating a scenario utilizing an automated scenario authoring system.

Scenario Authoring System – The automated tools required by a system scenario author to create/edit/delete/copy a scenario.

Scenario Generation – The translation of scenario author inputs and resident database information into the scenario-unique database information needed to execute the mission.

Search Radar – A radar whose prime function is to scan a specified volume of space and indicate the presence of any targets on some type of visual display and in some cases to provide coordinates of the target to a fire control system to assist in target acquisition and tracking.

Seeker – The seeker consists of circuitry in a homing missile which detects electronically examines and tracks the target. The seeker provides data for controlling the flight path of the missile and provides signals for destroying the missile or for detonating it at intercept.

Self-Screening Jamming (SSJ) – Each aircraft carries its own jamming equipment for its own protection.

Secondary Network – The collection of communication resources that provide connectivity among participants in an independent cluster (typically those units that are beyond connectivity range with the core instrumentation set).

Sensitivity – The sensitivity of a receiver is taken as the minimum signal level required to produce an output signal having a specified signal-to-noise ratio.

Sensor – The receiver portion of a transmitter and receiver pair used to detect and process electromagnetic energy.

Shipboard System – A Shipboard System is defined as a transportable ground/surface system that can be installed on a ship and can support deployed training operations.

Sideband – A signal either above or below the carrier frequency, produced by modulation of the carrier wave by some other wave.

Signal Strength – The magnitude of a signal at a particular location. Units are volts per meter

Single Point Failure – The failure of an item which would result in failure of the system and is not compensated for by redundancy or an alternative operating procedure.

Shop Replaceable Assembly (SRA)/Shop Replaceable Unit (SRU) – Naval Aviation/Naval Sea Systems Command terms (respectively) for the type of equipment for which Intermediate-level maintenance performs fault isolation and removal/replacement actions.

Simulated Engagement – The simulated launching of any weapon at a target.

Simulation – Synthetically representing the characteristics of a real world system or situation. In the trainer flight simulator context, the combat environment simulation represents selected characteristics of the behavior of

1. Physical or abstract entities including ships, aircraft, submarines, weapons, sensors, equipment.

2. The behavior of physical environmental characteristics or phenomena including weather, thermals, sea states.

3. Combat-related characteristics and events including command and control decisions and interactions, responses to various events, and tactics. All the above may encompass interactions with human operators, real tactical systems/equipment, and other simulated entities.

Simulation Fidelity – The degree with which a system or a portion of the system, accurately reproduces or corresponds to the essential characteristics, aspects, or elements of the simulated model and to the appropriate analogous characteristics of reality. This nominally represents the level or detail to which the model or simulation is developed and implemented. Adequacy for effective training is the primary factor in determination of the level of simulation fidelity required for a system. For example, if a weapon employment simulation produces sufficiently small percentages of false positives and false negatives to be no greater than random failures for real employment's of that weapon, then an adequate level of simulation fidelity for training would be achieved.

Spot Jamming – Narrow frequency band jamming concentrated against specific radar at a particular frequency.

The jamming bandwidth is comparable to the radar bandpass. The Spot Jammer can deny range and angle information.

Stand-Forward Jamming – A method which places the jamming vehicle between the enemy sensors and attack aircraft.

Stand-In Jamming – Similar to Stand-Forward Jamming but using an UAV with a lower powered jammer instead of a jammer aircraft.

Stand-Off Jamming (SOJ) – An ECM support aircraft orbits in the vicinity of the intended target. As the fighter-bomber pilot starts his strike penetration, the ECM aircraft directs jamming against all significant radar in the area.

The technique provides broad frequency band ECM without affecting performance of the strike aircraft.

Stimulation – The injection of artificial signals or data into an operational system to represent signals/data that are normally detected and/or processed by that system. In the system context, this definition includes analog signals, digital signals, data and message inputs into analog or digital equipment, systems, and data networks. Stimulation is the means by which software simulated combat environment characteristics or entities are "effected" or made real and known--"injected" or "inserted" into the combat system to promote interaction with the human operators and within the combat system itself.

Subassembly – Two or more parts which form a portion of an assembly or a unit replaceable as a whole, but having a part or parts which are individually replaceable.

Submarine – A manned sub-mersible vehicle. Subscriber node – An independent user of the communication service made available via the channel. Each generic subscriber node contains a data source, a data sink, a transmitter, a receiver, and a processor to coordinate the node's activities.

Subsurface platform – A submarine or an underwater weapon (torpedo or mine).

Support – A functional characterization of intent that implies assistance or helps that is not necessarily inherently required or immediately defined.

Suppression of Enemy Air Defenses (SEAD) – Activity which neutralizes, destroys, or temporarily degrades enemy air defense systems by using physical attack or electronic means.

Swept Jamming – Narrowband jamming which is swept through the desired frequency band in order to maximize power output. This technique is similar to spot noise jamming to create barrage jamming, but at a higher power.

Target Size – A measure of the ability of a radar target to reflect energy to the radar receiving antenna. The parameter used to describe this ability is the "radar cross section" of the target. The size of a target, such as an aircraft, will vary considerably as the target maneuvers and presents different views to the radar.

Terrain Bounce – Term for jamming that is directed at the earth's surface where it is reflected toward the threat radar.

Reflected jamming creates a virtual image of the jamming source on the earth as a target for HOJ missiles.

Threat – A track identified as hostile.

Track – A sequence of contacts made by a sensor system that is determined to represent a single entity. Tracks include platforms (aircraft, ships, and submarines) and employed weapons (missiles, torpedoes, etc.). Tracks have attributes such as position, altitude/depth, speed, heading, and category/classification.

Tracking Radar – A radar whose prime function is to track a radar target and determine the target coordinates so that a missile may be guided to the target or a gun aimed at the target.

Track While Scan Radar – The Track While Scan Radar provides complete and accurate position information for missile guidance. In one implementation it would utilize two separate beams produced by two separate antennas on two different frequencies. The system utilizes electronic computer techniques whereby raw datum is used to track an assigned target, compute target velocity and predict its future position, while maintaining normal sector scan.

Transponder – A transmitter/ receiver capable of accepting the electronic challenge of an interrogator and automatically transmitting an appropriate reply.

Velocity Gate Pull-Off (VGPO) – Method of capturing the velocity of a Doppler radar and moving it away from the skin echo. The VGPO is similar to the RGPO but used against CW or Doppler velocity tracking radar systems. The CW or pulse Doppler frequency is shifted in frequency to provide an apparent rate change or Doppler shift.

Viewpoint – The position and orientation of an observer in world coordinates.

Virtual World – The pseudo space that contains all simulated entities.

Wavelength – The distance traveled by a wave in one period, which is the time required to complete one cycle.

Window – Areas on the physical display in which it is possible to display graphical or textual data.

15.2 Acronyms

AAW	Anti Air Warfare
ACTS	Air Combat Training System
ADDS	Advanced Display and Debriefing Subsystem
ADL	Airborne Data Link
ADR	Arc Digitized Raster Graphics
AMRAAM	Advanced Medium Range Air-to-Air Missile
ADS	Air Defense System
AFB	Air Force Base
AFMSS	Air Force Mission Support System
AGL	Above Ground Level
AIS	Aircraft Instrumentation Subsystem
AISI	Aircraft Instrumentation Subsystem, Internal
AMW	Amphibious Warfare
AOC	Aerospace Operations Center
ARG	Amphibious Ready Group
ASC II	American Standard Code Information Interchange
ASUW	Anti Surface Warfare
ASW	Anti-Submarine Warfare
BFTT	Battle Force Tactical Trainer
BG	Battle Group
BIT	Built-In Test
BWS	Budgeted Work Schedule
C^3	Command, Control, and Communication
C^3I	Command, Control, Communication, and Intelligence
C^4I	Command, Control, Communications, Computer and Intelligence
C^4ISR	C^4I Surveillance and Reconnaissance
CCI	Controlled Cryptographic Item
CDRL	Contract Data Requirement Line
CGS	Computer-Generated Simulation
CGTS	Computer Generated Threat Systems
CI	Configuration Item

CM	Configuration Management
COE	Common Operating Environment
COTS	Commercial Off-the-Shelf
CP	Critical Path
CPU	Central Processor Unit
CPV	Critical Path Variant
CSC	Computer Software Components
CSCI	Computer Software Configuration Item
CSTS	Combat Simulation Test System
CSU	Computer Software Unit
CV	Aircraft Carrier
DFAD	Digital Feature Analysis Data
DII	Defense Information Infrastructure
DIICOE	Defense Information Infrastructure Common Operating Environment
DIS	Distributed Interactive Simulation
DOD	Department of Defense
DOP	Dilution of Precision
DLRP	Datalink Reference Point
DTED	Digital Terrain Elevation Data
ECE	Electronic Combat Environment
ECM	Electronic Countermeasures
ECEF	Earth-Centered, Earth-Fixed
EME	Electromagnetic Environment
EF	Expeditionary Force
EMC	Electromagnetic Compatibility
EME	Electromagnetic Environment
EMI	Electromagnetic Interference
EMV	Electromagnetic Vulnerability
EO	Electro-Optic
ERESS	Enhanced Radar Environmental Simulator System
ESD	Electrostatic Discharge
ESM	Electronic Support Measures
EUT	Equipment under Test
EW	Electronic Warfare
FAA	Federal Aviation Administration

FCA	Functional Configuration Audit
FINEX	Finish of Exercise
FTRG	Fleet Tactical Readiness Group
GFR	Gap-Filler Radar
GFS (1)	Government Furnished Software
GFS (2)	Government Furnished Simulation
GIM	Ground Interface Module
GPI	Ground Plane Interference
GPS	Global Positioning System
HDOP	Horizontal Dilution of Precision
HLA	High Level Architecture
HUD	Head-Up Display
HWCI	Hardware Configuration Item
IADS	Integrated Air Defense System
IBIT	Initiated Built-In Test
ICD	Interface Control Document
IDD	Interface Design Document
IDN	Instrumentation Datalink Network
IDT	Instrumentation Datalink Transceiver
IFF	Identification Friend or Foe
IMN	Indicated Mach Number
INS	Inertial Navigation System
I/O	Input/Output
IP	Internet Protocol
IR	Infra-Red
IWTS	Imaging Weapons Training System
JTA	Joint Technical Architecture
JTR	Joint Tactical Radio
JTCTS	Joint Tactical Combat Training System
LAN	Local Area Network
LFOCS	Live-Fire Operations Control Subsystem
LGB	Laser-Guided Bomb
LPD	Low Probability of Detection
LPI	Low Probability of Interception
LRU	Line Replaceable Unit
LTE	Launch to Detect

LSTS	Large Scale Target Sensing System
MIW	Mine Warfare
MRTS	Manned Radiating Simulators
MSL	Mean Sea Level
NAS	Naval Air Station
NEF	Naval Expeditionary Force
NERF	Naval Emitter Reference File
NDI	Non-Developmental Item
NDWS	No-Drop Weapon Scoring
NES	Network Encryption System
NMI	Nautical Mile
NOFORM	Non-Releaseable to Foreign Nationals
NTIA	National Telecommunications and Information Administration
NVLAP	National Voluntary Laboratory Accreditation Program
NWTDB	Naval Warfare Tactical Data Base
OBT	On-Board Trainer
OC^3	Orange Command Control and Communications
OFP	Operational Flight Program
OL	On-Line
OTAT	Over-the-Air Transfer
OTAR	Over-the Air Rekeying
PAT	Production Acceptance Testing
PBIT	Periodic Built-In Test
PDOP	Position Dilution of Precision
PIP	Participant Instrumentation Package
Pk	Probability of Kill
PMRF	Pacific Missile Range Facility
PRF	Pulse Repetition Frequency
RESS	Radar Environmental Simulator System
RF	Radio Frequency
RFMDS	Red Flag Measurement and Debriefing System
RRU	Remote Range Unit

RTKN	Real Time Kill Notification
RTS	Radiating Threat Simulator
SAI	Software Action Item
SDD	Software Design Document
SDL	SATCOM Datalink
SDP	Software Development Plan
SCORE	Southern California Offshore Range
SE	Support Equipment
SEAD	Suppression of Enemy Air Defenses
SEW	Space and Electronic Warfare
SIP	Ship Instrumentation Package
SLOC	Source (Software) Lines of Code
SRA	Shop-Replaceable Assembly
SRS	Software Requirements Specification
SRU	Shop-Replaceable Unit
STW	Strike Warfare
SVT	State-Vector Tracking
SWV	Schedule and Work Variance
TACTS	Tactical Aircrew Combat Training System
TAMPS	Tactical Air Mission Planning System
T&E	Test and Evaluation
TENA	Test and Training Enabling Architecture
TOT	Time-on-Target
TSPI	Time-Space-Position Information
UMTE	Unmanned Threat Emitter
USAF	U. S. Air Force
USMC	U. S. Marine Corps
USN	U. S. Navy
UTC	Universal Time Clock
VDD	Version Description Document
VDOP	Vertical Dilution of Precision
VRMS	Volt Root Mean Square
WAN	Wide Area Network
WP	Work Performed
WRA	Weapon Replaceable Assembly
WSSA	Weapon Systems Support Activity

15.3 Index

K

L

M

N

Q

R

S

W

X

Y

Z

Appendix A Engineers, Military and Managers Working with G. Terry Thomas

Abel, Jim	Naval Engineer
Abrams, Don	Veridian Company
Alfonsi, Eric	Naval Engineer
Alicia, Shawn	Naval Engineer
Alessandro, Frank	Naval Engineer
Ashenfelter, Andrew	Naval Engineer
Atay, Alpuan	Naval Manager
Bangs, Mary	Naval Engineer
Baker, George	Veridian Company
Barstow, John	Eagan McAllister Company
Baum, Barbara	Naval Engineer
Becker, Bob	Naval Manager
Bergey, John	Naval Manager
Bischoff, David	Veridian Company
Boone, Pepper	Porter Technology Company
Bouchard, Bill	Eagan McAllister Company
Brickman, Philip	Naval Engineer
Bromberger, Dick	Naval Engineer
Buscemi, Tony	Naval Engineer
Butkus, Richard	Naval Manager
Calahan, Pat	Naval Manager
Carmack, Dale	Naval Manager
Chamber, Walt	Naval Manager
Cherry, Vic	Naval Engineer
Christian, Whitey	Naval Manager
Cicero, Glenn	Naval Engineer
Chin, Bob	Naval Engineer
Clare, Warren	Naval Engineer
Columbo, Joseph	Naval Manager
Croteau, Rudolph	Naval Engineer

Crosbie, Richard	Naval Manager
Crosby, Peter	Naval Engineer
Dafrico, Richard	Naval Engineer
Davis, Jeff	Naval Engineer
Davis, Scott	Naval Engineer
DeCrescente, Carmine	Naval Manager
Delisi George	Naval Engineer
Dias, Hector	Naval Engineer
DiCola, John	Naval Manager
Dolceamore, Robert	Naval Manager
Drobeck, Tom	Naval Manager
Dyer, Radm Joe	Naval Officer
Eberz, Carol	Naval Support
Eggen, Dale	Veridan Company
Eyth, Jack	Naval Manager
Feeman, Maj Bob	Air Force Officer
Feeser, Tony	Naval Engineer
Fesnak, Alan	Naval Engineer
Fisher, Capt Rory	Naval Officer
Franklin, Rich	Dimensions International Company
Furlin, Roger	Naval Manager
Galloway, Tom	Naval Engineer
Geist, Karl	Naval Engineer
Gibb, Stewart	Naval Manager
Gill, Janet	Naval Engineer
Ginn, Robert	Naval Engineer
Glemser, Ray	Naval Engineer
Goodman, Rick	Grumman Aircraft Company
Guarini, Jerry	Naval Manager
Guido, Anthony	Naval Manager
Gum, Donald	Air Force Manager
Gozzo, Lt Cdr Steve	Naval Officer
Gramp, Al	Naval Engineer
Halikman, Edward	A&E Company
Hall, Cdr Douglas	Naval Officer
Hartman, Richard	McFadden Company

Harvey, Ed	BMH Company
Harvey, Ken	Naval Manager
Heffner, Peggy	Naval Engineer
Hendricks, Steve	Naval Manager
Hester, John	Naval Engineer
Hicks, Brian	Naval Manager
Hitchcock, Lloyd	Naval Engineer
Hoffman, Jim	Naval Engineer
Hribar, Adrien	Veridian Company
Hummel, Darryl	Veridian Company
Hungerford, Ted	Naval Engineer
Ivery, Barbette	Naval Engineer
Jankiewicz, Bob	Naval Engineer
Jefferies, Kent	AMEWAS Company
Jeffries, Amy	AMEWAS Company
Johansen, Joey	Naval Engineer
Jones, Bob	Naval Manager
Jordan, Michael	Grumman Aircraft Company
Kaniss, Al	Naval Manager
Kermon, Lawrence	MicroSim Company
Kieffer, Dennis	Naval Engineer
Kilchenmann, Mark	Holmes Tucker Company
Keller, Jim	Naval Manager
Knouse, Barry	Naval Engineer
Kolb, Michael	Naval Manager
Koper, Harry	Naval Engineer
Kori, Suresh	Naval Manager
Kuhn, Alexander	Naval Manager
Lamoreaux, Cdr Scott	Naval Officer
Lee, Tony	Naval Engineer
Lewandowski, Dean	Air Force Engineer
Lindenberg, Klaus	Enzian Technology Company
Lizbinski, Michael	Naval Engineer
Long, Gordon	Naval Engineer
Magnus, Ralph	Information Spectrum Company
Mahoney, Brian	Veridan Company

McDonagh, Dan	Naval Engineer
McFadden, Burt	McFadden Company
McTigue, Kevin	Virual Technology Company
McCombs, Cheryl	Naval Engineer
McCurdy, Jesse	Naval Manager
Mercer, Lt Cdr Mike	Naval Officer
Miller, Jerry	Naval Engineer
Mirales, Nick	Naval Manager
Mitchell, Dick	Naval Manager
Morello, Major Kevin	Air Force Officer
Morrison, Dan	Lockheed Martin Company
Mudryk, John	Naval Engineer
Murnin, Robert	Naval Engineer
Murphy, Cdr Skip	Naval Officer
Murray, Tom	Naval Engineer
Mutschler, David	Naval Engineer
Nave, Ron	Naval Engineer
Odgen, Bill	Naval Manager
Ostroff, Norm	Naval Engineer
Peron, James	Naval Engineer
Patterson, Dan	Naval Engineer
Payret, Charlie	Grumman Aircraft Company
Pepka, Ron	Naval Contracts
Petrone, Vincent	Naval Engineer
Philips, John	Naval Engineer
Piranian, Al	Naval Engineer
Pohle, Bill	Naval Engineer
Poore, Lee	Naval Engineer
Przybylowski, Joe	Naval Manager
Quade, Lisa	Eagan McAllister Company
Raney, Dale	Naval Engineer
Reed, Edgar	Naval Manager
Riner, Bruce	Naval Engineer
Risko, Theodore	Naval Engineer
Roschetz, Lynn	Naval Engineer
Rosenthal, Steve	Naval Engineer

Rovinsky, Richard	Air Force Engineer
Russo, Lou	Naval Engineer
Sabatini, Carolyn	Naval Engineer
Samtmann, Robert	Naval Engineer
Santangelo, Tom	Naval Engineer
Santini, John	Naval Manager
Sarkovitz, Mike	Naval Manager
Schab, Dan	Naval Engineer
Schaff, Joe	Naval Engineer
Schmidt, Ed	Naval Engineer
Schott, David	DMOC Company
Schumacher, Denny	Information Spectrum, Inc
Schwebke, Russ	TACCSF Company
Shea, John	BGI Company
Sieffert, William	Naval Manager
Shocket, Fred	Naval Engineer
Singmore, Virginia	Naval Contracts
Skullman, Vic	Naval Engineer
Stachelczyk, Gregory	Information Spectrum Company
Svecz, Mike	Naval Engineer
Tafel, Bob	Naval Manager
Tidwell, Jim	Naval Engineer
Teper, Gary	Information Spectrum, Inc
Thomas, Brian	Naval Engineer
Toth, Jim	Naval Manager
Tseng, Jine	Naval Manager
Yannayon, Nancy	AMEWAS Company
Varma, Asha	Naval Engineer
Walker, John	Naval Manager
Webster, Charles	Naval Engineer
Westerbeke, Cindy	Naval Manager
Westerhouse, Steve	McDonnell Douglas Company
Williams, Dave	Naval Engineer
Williamson, Tony	Naval Engineer
Wolfe, Norman	Aver Star Company
Wolter, Paul	Naval Engineer

Engineers and Managers Working with G. Terry Thomas

Wong, Ying	Naval Engineer
Wentz, Bill	Naval Engineer
Zettler, Maynard	Naval Manager
Zwissler, Robert	Naval Engineer

Appendix B Systems Engineering Mathematical Terms Presented below is some general mathematical Information taken from a Pt Mugu Naval Base Report.

CONSTANTS, CONVERSIONS, and CHARACTERS

DECIMAL MULTIPLIER PREFIXES

Prefix	Symbol	Multiplier
exa	E	10^{18}
peta	P	10^{15}
tera	T	10^{12}
giga	G	10^{9}
mega	M	10^{6}
kilo	k	10^{3}
hecto	h	10^{2}
deka	da	10^{1}
deci	d	10^{-1}
centi	c	10^{-2}
milli	m	10^{-3}
micro	μ	10^{-6}
nano	n	10^{-9}
pico	p	10^{-12}
femto	f	10^{-15}
atto	a	10^{-18}

EQUIVALENCY SYMBOLS

Symbol	Meaning
$\propto$	Proportional
$\sim$	Roughly equivalent
$\approx$	Approximately
$\cong$	Nearly equal
$=$	Equal
$\equiv$	Identical to, defined as
$\neq$	Not equal
$>>$	Much greater than
$>$	Greater than
$\geq$	Greater than or equal to
$<<$	Much less than
$<$	Less than
$\leq$	Less than or equal to
$\therefore$	Therefore
$\circ$	Degrees
$'$	Minutes or feet
$''$	Seconds or inches

UNITS OF LENGTH

1 inch (in)	=	2.54 centimeters (cm)
1 foot (ft)	=	30.48 cm = 0.3048 m
1 yard (yd)	$\cong$	0.9144 meter
1 meter (m)	$\cong$	39.37 inches
1 kilometer (km)	$\cong$	0.54 nautical mile
	$\cong$	0.62 statute mile
	$\cong$	1093.6 yards
	$\cong$	3280.8 feet
1 statute mile	$\cong$	0.87 nautical mile
(sm or stat. mile)	$\cong$	1.61 kilometers
	=	1760 yards
	=	5280 feet
1 nautical mile	$\cong$	1.15 statute miles
(nm or naut. mile)	$\cong$	1.852 kilometers
	$\cong$	2025 yards
	$\cong$	6076 feet
1 furlong	=	1/8 mi (220 yds)

UNITS OF SPEED

1 foot/sec (fps)	$\cong$	0.59 knot (kt)*
	$\cong$	0.68 stat. mph
	$\cong$	1.1 kilometers/hr
1000 fps	=	600 knots
1 kilometer/hr	$\cong$	0.54 knot
(km/hr)	$\cong$	0.62 stat. mph
	$\cong$	0.91 ft/sec
1 mile/hr (stat.)	$\cong$	0.87 knot
(mph)	$\cong$	1.61 kilometers/hr
	$\cong$	1.47 ft/sec
1 knot*	$\cong$	1.15 stat. mph
	$\cong$	1.69 feet/sec
	$\cong$	1.85 kilometer/hr
	$\cong$	0.515 m/sec

*A knot is 1 nautical mile per hour.

UNITS OF VOLUME

1 gallon	≅	3.78 liters
	≅	231 cubic inches
	≅	0.1335 cubic ft
	≅	4 quarts
	≅	8 pints
1 fl ounce	≅	29.57 cubic centimeter (cc) or milliliters (ml)
1 in^3	≅	16.387 cc

UNITS OF AREA

1 sq meter	≅	10.76 sq ft
1 sq in	≅	645 sq millimeters (mm)
	=	1,000,000 sq mil
1 mil	=	0.001 inch
1 acre	=	43,560 sq ft

UNITS OF WEIGHT

1 kilogram (kg)	≅	2.2 pounds (lbs)
1 pound	≅	0.45 Kg
	=	16 ounce (oz)
1 oz	=	437.5 grains
1 carat	≅	200 mg
1 stone (U.K.)	≅	6.36 kg

NOTE: These are the U.S. customary (avoirdupois) equivalents, the troy or apothecary system of equivalents, which differ markedly, was used long ago by pharmacists.

UNITS OF POWER / ENERGY

1 H.P.	=	33,000 ft-lbs/min
	=	550 ft-lbs/sec
	≅	746 Watts
	≅	2,545 BTU/hr

(BTU = British Thermal Unit)

1 BTU	≅	1055 Joules
	≅	778 ft-lbs
	≅	0.293 Watt-hrs

SCALES
OCTAVES

"N" Octaves = Freq to Freq x 2^N
i.e. One octave would be 2 to 4 GHz
Two Octaves would be 2 to 8 GHz
Three octaves would be 2 to 16 GHz

DECADES

"N" Decades = Freq to Freq x 10^N
i.e. One decade would be 1 to 10 MHz
Two decades would be 1 to 100 MHz
Three decades would be 1 to 1000 MHz

TEMPERATURE CONVERSIONS

$°F = (9/5)°C + 32$

$°C = (5/9)(°F - 32)$

$°K = °C + 273.16$

$°F = (9/5)(°K - 273) + 32$

$°C = °K - 273.16$

$°K = (5/9)(°F - 32) + 273$

UNITS OF TIME

1 year	=	365.2 days
1 fortnight	=	14 nights (2 weeks)
1 century	=	100 years
1 millennium	=	1,000 years

NUMBERS

1 decade	=	10
1 Score	=	20
1 Billion	=	1×10^9 (U.S.) (thousand million)
	=	1×10^{12} (U.K.)

RULE OF THUMB FOR ESTIMATING DISTANCE TO LIGHTNING / EXPLOSION:

km - Divide 3 into the number of seconds which have elapsed between seeing the flash and hearing the noise.
miles - Multiply 0.2 times the number of seconds which have elapsed between seeing the flash and hearing the noise.

Note: Sound vibrations cause a change of density and pressure within a media, while electromagnetic waves do not. An audio tone won't travel through a vacuum but can travel at 1100 ft/sec through air. When picked up by a microphone and used to modulate an EM signal, the modulation will travel at the speed of light.

Physical Constant	Quoted Value	S*	SI unit	Symbol
Avogadro constant	6.0221367×10^{23}	36	mol^{-1}	N_A
Bohr magneton	$9.2740154 \times 10^{-24}$	31	$J \cdot T^{-1}$	μ_B
Boltzmann constant	1.380658×10^{-23}	12	$J \cdot K^{-1}$	$k(=R\,N_A)$
Electron charge	$1.602177\,33 \times 10^{-19}$	49	C	-e
Electron specific charge	$-1.758819\,62 \times 10^{11}$	53	$C \cdot kg^{-1}$	$-e/m_e$
Electron rest mass	$9.1093897 \times 10^{-31}$	54	kg	m_e
Faraday constant	9.6485309×10^4	29	$C \cdot mol^{-1}$	F
Gravity (Standard Acceleration)	9.80665 or 32.174	0	m/sec^2 ft/sec^2	g
Josephson frequency to voltage ratio	4.8359767×10^{14}	0	$Hz \cdot V^{-1}$	2e/hg
Magnetic flux quantum	$2.06783461 \times 10^{-15}$	61	Wb	ϕ_o
Molar gas constant	8.314510	70	$J \cdot mol^{-1} \cdot K^{-1}$	R
Natural logarithm base	$\cong 2.71828$	-	dimensionless	e
Newtonian gravitational constant	6.67259×10^{-11}	85	$m^3 \cdot kg^{-1} \cdot s^{-2}$	G or K
Permeability of vacuum	$4\pi \times 10^{-7}$	d	H/m	μ_o
Permittivity of vacuum	$\approx 8.8541878 \times 10^{-12}$	d	F/m	ϵ_o
Pi	$\cong 3.141592654$		dimensionless	π
Planck constant	6.62659×10^{-34}	40	$J \cdot s$	h
Planck constant/2π	$1.05457266 \times 10^{-34}$	63	$J \cdot s$	$h(=h2\pi)$
Quantum of circulation	$3.63694807 \times 10^{-4}$	33	$J \cdot s \cdot kg^{-1}$	$h/2m_e$
Radius of earth (Equatorial)	6.378×10^6 or 3963		m miles	
Rydberg constant	1.0973731534×10^7	13	m^{-1}	R_y
Speed of light	2.9979246×10^8	1	$m \cdot s^{-1}$	c
Speed of sound (dry air @ std press & temp)	331.4	-	$m \cdot s^{-1}$	-
Standard volume of ideal gas	22.41410×10^{-3}	19	$m^3 \cdot mol^{-1}$	V_m
Stefan-Boltzmann constant	5.67051×10^{-8}	19	$W \cdot K^{-4} \cdot m^{-2}$	σ

* S is the one-standard-deviation uncertainty in the last units of the value, d is a defined value.
(A standard deviation is the square root of the mean of the sum of the squares of the possible deviations)

GREEK ALPHABET

Case Upper	Case Lower	Greek Alphabet Name	English Equivalent	Case Upper	Case Lower	Greek Alphabet Name	English Equivalent
A	α	alpha	a	N	ν	nu	n
B	β	beta	b	Ξ	ξ	xi	x
Γ	γ	gamma	g	O	o	omicron	ŏ
Δ	δ	delta	d	Π	π	pi	p
E	ε	epsilon	ĕ	P	ρ	rho	r
Z	ζ	zeta	z	Σ	σ	sigma	s
H	η	eta	ē	T	τ	tau	t
Θ	θ, ϑ	theta	th	Υ	υ	upsilon	u
I	ι	iota	i	Φ	φ, φ	phi	ph
K	κ	kappa	k	X	χ	chi	ch
Λ	λ	lambda	l	Ψ	ψ	psi	ps
M	μ	mu	m	Ω	ω	omega	ō

LETTERS FROM THE GREEK ALPHABET COMMONLY USED AS SYMBOLS

Symbol	Name	Use
α	alpha	space loss, angular acceleration, or absorptance
β	beta	3 dB bandwidth or angular field of view [radians]
Γ	Gamma	reflection coefficient
γ	gamma	electric conductivity, surface tension, missile velocity vector angle, or gamma ray
Δ	Delta	small change or difference
δ	delta	delay, control forces and moments applied to missile, or phase angle
ε	epsilon	emissivity [dielectric constant] or permittivity [farads/meter]
η	eta	efficiency or antenna aperture efficiency
Θ	Theta	angle of lead or lag between current and voltage
θ or ϑ	theta	azimuth angle, bank angle, or angular displacement
Λ	Lambda	acoustic wavelength or rate of energy loss from a thermocouple
λ	lambda	wavelength or Poisson Load Factor
μ	mu	micro 10^{-6} [micron], permeability [henrys/meter], or extinction coefficient [optical region]
ν	nu	frequency
π	pi	3.141592654+
ρ	rho	charge/mass density, resistivity [ohm-meter], VSWR, or reflectance
Σ	Sigma	algebraic sum
σ	sigma	radar cross section [RCS], Conductivity [1/ohm-meter], or Stefan-Boltzmann constant
T	Tau	VSWR reflection coefficient
τ	tau	pulse width, atmospheric transmission, or torque
Φ	Phi	magnetic/electrical flux, radiant power [optical region], or Wavelet's smooth function [low pass filter]
φ or φ	phi	phase angle, angle of bank, or beam divergence [optical region]
Ψ	Psi	time-dependent wave function or Wavelet's detail function [high pass filter]
ψ	psi	time-independent wave function, phase change, or flux linkage [weber]
Ω	Omega	Ohms [resistance] or solid angle [optical region]. Note: inverted symbol is conductance [mhos]
ω	omega	carrier frequency in radians per second

MORSE CODE and PHONETIC ALPHABET

A - alpha	• -	J - juliett	• - - -	S - sierra	• • •	1	• - - - -
B - bravo	- • • •	K - kilo	- • -	T - tango	-	2	• • - - -
C - charlie	- • - •	L - lima	• - • •	U - uniform	• • -	3	• • • - -
D - delta	- • •	M - mike	- -	V - victor	• • • -	4	• • • • -
E - echo	•	N - november	- •	W - whiskey	• - -	5	• • • • •
F - foxtrot	• • - •	O - oscar	- - -	X - x-ray	- • • -	6	- • • • •
G - golf	- - •	P - papa	• - - •	Y - yankee	- • - -	7	- - • • •
H - hotel	• • • •	Q - quebec	- - • -	Z - zulu	- - • •	8	- - - • •
I - india	• •	R - romeo	• - •	0	- - - - -	9	- - - - •

Note: The International Maritime Organization agreed to officially stop Morse code use by February 1999, however use may continue by ground based amateur radio operators (The U.S. Coast Guard discontinued its use in 1995).

Basic Math / Geometry Review

EXPONENTS

$$a^x a^y = a^{x+y}$$

$$a^x / a^y = a^{x-y}$$

$$(a^x)^y = a^{xy}$$

$$a^0 = 1$$

Example:

$$\frac{x}{\sqrt{x}} = x \cdot x^{-\frac{1}{2}} = x^{(1-\frac{1}{2})} = x^{\frac{1}{2}} = \sqrt{x}$$

LOGARITHMS

$$\log (xy) = \log x + \log y$$

$$\log (x/y) = \log x - \log y$$

$$\log (x^N) = N \log x$$

If $z = \log x$ then $x = 10^z$

Examples: $\log 1 = 0$
$\log 1.26 = 0.1$; $\log 10 = 1$

if $10 \log N = dB\#$,
then $10^{(dB\#/10)} = N$

TRIGONOMETRIC FUNCTIONS

$$\sin x = \cos (x-90°)$$

$$\cos x = -\sin (x-90°)$$

$$\tan x = \sin x / \cos x = 1 / \cot x$$

$$\sin^2 x + \cos^2 x = 1$$

A radian is the angular measurement of an arc which has an arc length equal to the radius of the given circle, therefore there are 2π radians in a circle. One radian = $360°/2\pi = 57.296....°$

ELLIPSE

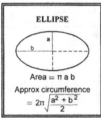

Area $= \pi\, a\, b$
Approx circumference
$= 2\pi \sqrt{\dfrac{a^2 + b^2}{2}}$

RECTANGLULAR SOLID

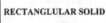

Area $=$ lw
Volume $=$ lwh

CYLINDER

Volume $=$ πr2h
Lateral surface
area $=$ 2πrh

ANGLES

Sin θ $=$ y/r   Cos θ $=$ x/r
Tan θ $=$ y/x   r2 $=$ x2 + y 2

THE SPEED OF LIGHT			
ACTUAL	UNITS	RULE OF THUMB	UNITS
$\cong 2.9979246 \times 10^8$	m/sec	$\approx 3 \times 10^8$	m/sec
$\cong 299.79$	m/µsec	≈ 300	m/µsec
$\cong 3.27857 \times 10^8$	yd/sec	$\approx 3.28 \times 10^8$	yd/sec
$\cong 5.8275 \times 10^8$	NM/hr	$\approx 5.8 \times 10^8$	NM/hr
$\cong 1.61875 \times 10^5$	NM/sec	$\approx 1.62 \times 10^5$	NM/sec
$\cong 9.8357105 \times 10^8$	ft/sec	$\approx 1 \times 10^9$	ft/sec

SPEED OF LIGHT IN VARIOUS MEDIUMS

The speed of EM radiation through a substance such as cables is defined by the following formula:

$$V = c/(\mu_r \epsilon_r)^{1/2}$$

Where: μ_r = relative permeability
 ϵ_r = relative permittivity
The real component of ϵ_r = dielectric constant of medium.

EM propagation speed in a typical cable might be 65-90% of the speed of light in a vacuum.

APPROXIMATE SPEED OF SOUND (MACH 1)

Sea Level (CAS/TAS)		36,000 ft* (TAS)	(CAS)
1230 km/hr	Decreases	1062 km/hr	630 km/hr
765 mph	Linearly	660 mph	391 mph
665 kts	To ⇝	573 kts	340 kts

* The speed remains constant until 82,000 ft, when it increases linearly to 1215 km/hr (755 mph, 656 kts) at 154,000 ft. Also see section 8-2 for discussion of Calibrated Air Speed (CAS) and True Airspeed (TAS) and a plot of the speed of sound vs altitude.

SPEED OF SOUND IN VARIOUS MEDIUMS

Substance	Speed (ft/sec)
Vacuum	Zero
Air	1,100
Fresh Water	4,700
Salt Water	4,900
Glass	14,800

DECIMAL / BINARY / HEX CONVERSION TABLE

Decimal	Binary	Hex	Decimal	Binary	Hex	Decimal	Binary	Hex
1	00001	01h	11	01011	0Bh	21	10101	15h
2	00010	02h	12	01100	0Ch	22	10110	16h
3	00011	03h	13	01101	0Dh	23	10111	17h
4	00100	04h	14	01110	0Eh	24	11000	18h
5	00101	05h	15	01111	0Fh	25	11001	19h
6	00110	06h	16	10000	10h	26	11010	1Ah
7	00111	07h	17	10001	11h	27	11011	1Bh
8	01000	08h	18	10010	12h	28	11100	1Ch
9	01001	09h	19	10011	13h	29	11101	1Dh
10	01010	0Ah	20	10100	14h	30	11110	1Eh

When using hex numbers it is always a good idea to use "h" as a suffix to avoid confusion with decimal numbers.

To convert a decimal number above 16 to hex, divide the number by 16, then record the integer resultant and the remainder. Convert the remainder to hex and write this down - this will become the far right digit of the final hex number. Divide the integer you obtained by 16, and again record the new integer result and new remainder. Convert the remainder to hex and write it just to the left of the first decoded number. Keep repeating this process until dividing results in only a remainder. This will become the left-most character in the hex number. i.e. to convert 60 (decimal) to hex we have 60/16 = 3 with 12 remainder. 12 is C (hex) - this becomes the right most character. Then 3/16=0 with 3 remainder. 3 is 3 (hex). This becomes the next (and final) character to the left in the hex number, so the answer is 3C.

Systems Engineering Mathematical Terms Presented

TRIANGLES

Angles: $A + B + C = 180°$

$c^2 = a^2 + b^2 - 2ab \cos C$

Area $= 1/2 \, bh = 1/2 \, ac \sin B$

$c = \sqrt{d^2 + h^2}$

SPHERE

Surface area $= 4\pi r^2$

Volume $= 4/3 \, \pi r^3$

Cross Section (circle)
Area $= \pi r^2$

Circumference (c) $= 2\pi r$

DERIVATIVES

Assume: a = fixed real #; u, v & w are functions of x

$d(a)/dx = 0$; $d(\sin u)/dx = du(\cos u)/dx$

$d(x)/dx = 1$; $d(\cos v)/dx = -dv(\sin v)/dx$

$d(uvw)/dx = uvdw/dx + vwdu/dx + uwdv/dx +...etc$

INTEGRALS

Note: All integrals should have a constant of integration added

Assume: a = fixed real #; u, & v are functions of x

$\int a \, dx = ax$ and $\int a \, f(x)dx = a\int f(x)dx$

$\int (u + v)dx = \int u \, dx + \int v \, dx$; $\int e^x dx = e^x$

$\int (\sin ax)dx = -(\cos ax)/a$; $\int (\cos ax)dx = (\sin ax)/a$

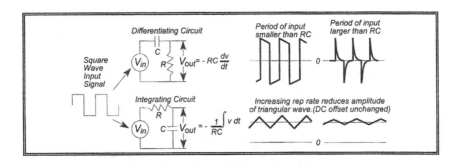

FREQUENCY SPECTRUM

Figure 1, which follows, depicts the electromagnetic radiation spectrum and some of the commonly used or known areas. Figure 2 depicts the more common uses of the microwave spectrum. Figure 3 shows areas of the spectrum which are frequently referred to by band designations rather than by frequency.

Section 7-1 provides an additional breakdown of the EO/IR spectrum.

To convert from frequency (f) to wavelength (λ) and vice versa, recall that $f = c/\lambda$, or $\lambda = c/f$; where c = speed of light.

$$\lambda_{meter} = \frac{3x10^8}{f_{Hz}} = \frac{3x10^5}{f_{kHz}} = \frac{300}{f_{MHz}} = \frac{0.3}{f_{GHz}} \qquad \text{or} \qquad f_{Hz} = \frac{3x10^8}{\lambda_{meter}} \quad f_{kHz} = \frac{3x10^5}{\lambda_{meter}} \quad f_{MHz} = \frac{300}{\lambda_{meter}} \quad f_{GHz} = \frac{0.3}{\lambda_{meter}}$$

Some quick rules of thumb follow:

Metric:

        Wavelength in cm = 30 / frequency in GHz
        For example: at 10 GHz, the wavelength = 30/10 = 3 cm

English:

        Wavelength in ft = 1 / frequency in GHz
        For example: at 10 GHz, the wavelength = 1/10 = 0.1 ft

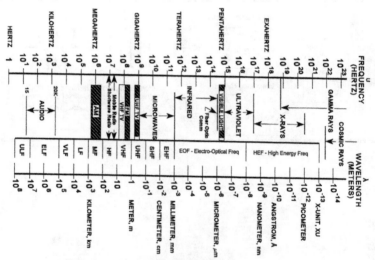

Electromagnetic Radiation Spectrum

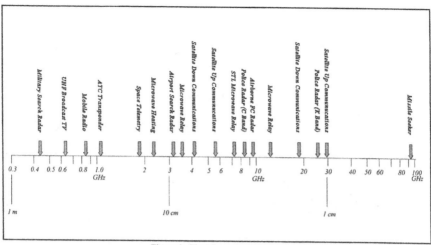

Figure 2. The Microwave Spectrum

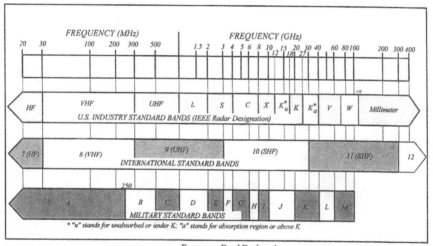

Frequency Band Designations